AF734556

INTERNET THEORY, TECHNOLOGY AND APPLICATIONS

AMERICA'S EMBRACE OF THE MOBILE INTERNET

ANALYSES AND ISSUES

Internet Theory, Technology and Applications

Additional books in this series can be found on Nova's website under the Series tab.

Additional e-books in this series can be found on Nova's website under the e-book tab.

INTERNET THEORY, TECHNOLOGY AND APPLICATIONS

AMERICA'S EMBRACE OF THE MOBILE INTERNET

ANALYSES AND ISSUES

JENNIFER HAGGARD
EDITOR

nova publishers
New York

Copyright © 2015 by Nova Science Publishers, Inc.

All rights reserved. No part of this book may be reproduced, stored in a retrieval system or transmitted in any form or by any means: electronic, electrostatic, magnetic, tape, mechanical photocopying, recording or otherwise without the written permission of the Publisher.

For permission to use material from this book please contact us:
Telephone 631-231-7269; Fax 631-231-8175
Web Site: http://www.novapublishers.com

NOTICE TO THE READER

The Publisher has taken reasonable care in the preparation of this book, but makes no expressed or implied warranty of any kind and assumes no responsibility for any errors or omissions. No liability is assumed for incidental or consequential damages in connection with or arising out of information contained in this book. The Publisher shall not be liable for any special, consequential, or exemplary damages resulting, in whole or in part, from the readers' use of, or reliance upon, this material. Any parts of this book based on government reports are so indicated and copyright is claimed for those parts to the extent applicable to compilations of such works.

Independent verification should be sought for any data, advice or recommendations contained in this book. In addition, no responsibility is assumed by the publisher for any injury and/or damage to persons or property arising from any methods, products, instructions, ideas or otherwise contained in this publication.

This publication is designed to provide accurate and authoritative information with regard to the subject matter covered herein. It is sold with the clear understanding that the Publisher is not engaged in rendering legal or any other professional services. If legal or any other expert assistance is required, the services of a competent person should be sought. FROM A DECLARATION OF PARTICIPANTS JOINTLY ADOPTED BY A COMMITTEE OF THE AMERICAN BAR ASSOCIATION AND A COMMITTEE OF PUBLISHERS.

Additional color graphics may be available in the e-book version of this book.

Library of Congress Cataloging-in-Publication Data

ISBN: 978-1-63463-585-1

Published by Nova Science Publishers, Inc. † New York

CONTENTS

PREFACE

The rise of sophisticated mobile devices and high-speed wireless networks is transforming the Internet and how our society interacts with it. Americans are rapidly embracing mobile technologies and the opportunities they create. This book examines data from the 2012 Current Population Survey (CPS) describing Americans' use of computers and the Internet. Data from this survey depict the accelerating popularity of mobile phone-based Internet use across demographic groups. The book also reveals the explosion in mobile Internet use accompanied by narrowing but persistent disparities in computer ownership and Internet use overall. This book's insights will allow policymakers, in cooperation with private partners, to make decisions that will improve America's productivity, ingenuity, world competitiveness, and standard of living.

Chapter 1 – *Exploring the Digital Nation: Embracing the Mobile Internet* builds on the last *Digital Nation* study (NTIA & ESA, 2013) and examines data from the 2012 Current Population Survey (CPS) describing Americans' use of computers and the Internet. Data from this survey depict the accelerating popularity of mobile phone-based Internet use across American demographic groups. The Census Bureau collected data from over 53,000 households in the October 2012 CPS Computer and Internet Use Supplement, forming a representative sample of the 122 million households across the United States. NTIA's analysis of this dataset, summarized below, reveals a mobile Internet explosion, accompanied by narrowing but persistent disparities in computer ownership and Internet use overall.

As in all installments of *Digital Nation*, this report includes NTIA's analysis of the demographic characteristics of home Internet users and the technologies they use to go online, as well as the alternative locations where they use the Internet. The report also presents data on the primary reasons why some Americans do not access the Internet from their homes. These data can inform further research and policy that strives to eliminate disparities by increasing the availability of affordable broadband in the United States and promoting broadband adoption by all Americans.

Chapter 2 – The use of mobile devices for communications and information exchange has been shown to be very desirable, with more than 5.3 billion mobile subscribers estimated (mobiThinking, 2011) and more than half-billion mobile phones sold each year (Blom, Chipchase, and Lehikoinen, 2005). Most often, when the term "mobile devices" is used, the authors really mean "mobile phones." For example, Blom et al. (2005) describes "personal, mobile synchronous and asynchronous communication for all types of users..." (p. 37) and then cites that more than one half a billion mobile phones are sold each year. However, mobile devices are not limited to phones; in addition to cellular phones, there are also personal digital assistants (e.g., Blackberry) as well as the more recent advent of wireless tablet computers (e.g., iPad). With these "smart" mobile devices, users can communicate by voice and text as well as have Internet access as a means of communications and information exchange. Mobile device use has grown tremendously; for example, in 2002, mobile (cellular) phones exceeded the number of landlines (Donner, 2008). All types of mobile devices have increased as well; for example, it is estimated that 25 million iPads have been sold to date (Rosoff, 2011).

The purpose of this report is to explore some of the research literature that discusses mobile technology, specifically mobile phones and mobile access to the Internet, and the impact that culture has on its adoption, use and design. One of the results of this review will be to identify factors that are important when considering mobile technology and culture, specifically, and factors that may be important in examining the impact of culture on decision making and communication, more generally. To accomplish this goal, the report first looks at culture, its definition, models of culture and dimensions used to describe specific cultures. The next section then examines mobile technology and culture, focusing on (1) models of adoption of technology, specifically mobile technology; (2) examples of uses of mobile technology and the influence of culture, and (3) design of mobile user interfaces and experience considering cultural influences. Finally, the authors discuss the information identified and its relevance to their goals of building a cognitive framework for social-cultural modeling, as well as other observations concerning technology and culture, especially military applications.

Chapter 3 – The first text messages were sent during 1992 and 1993, although commercially, text messaging was not widely offered or used until 2000. Even then, messages could only be sent between users subscribed to the same wireless carrier; for example, Sprint customers could only exchange messages with other Sprint customers. In November 2001, however, wireless service providers began to connect their networks for text messaging, allowing subscribers on different networks to exchange text messages. Since then, the number of text messages in the United States has grown to over 48 billion messages every month. Additionally, text messages are no longer only sent as "point-to-point" communications between two mobile device users. More specifically, messages are also commonly sent from web-based applications within a web browser (e.g., from an Internet e-mail address) and from instant messaging clients like AIM or MSN.

For congressional policymakers, two major categories of issues have arisen: (1) "same problem, different platform" and (2) issues stemming from the difficulty in applying existing technical definitions to a new service, such as whether a text message is sent "phone-to-phone" or using the phone's associated email address. There are numerous examples of each. An example of the first category would be consumer fraud and children's accessing inappropriate content, which have existed previously in the "wired world," but have now found their way to the "wireless world." An example of the second category would be that spam sent between two phones or from one phone to many phones does not fall under the definition of spam in the CAN-SPAM Act of 2003 (Controlling the Assault of Non-Solicited Pornography and Marketing Act, P.L. 108-187); however, if that same message were to be sent from a phone or computer using the phone's associated e-mail address, it would.

The increasing use of text and multimedia messaging has raised several policy issues: distracted driving, SMS spam, the inability of consumers to disable text messaging, text messaging price fixing, carrier blocking of common short code messages, deceptive and misleading common short code programs, protecting children from inappropriate content on wireless devices, "sexting," mobile cyberbullying, privacy of text messages, and using SMS to support law enforcement and emergency response.

In: America's Embrace of the Mobile Internet ISBN: 978-1-63463-585-1
Editor: Jennifer Haggard © 2015 Nova Science Publishers, Inc.

Chapter 1

EXPLORING THE DIGITAL NATION: EMBRACING THE MOBILE INTERNET*

U.S. Department of Commerce, National Telecommunications and Information Administration

FOREWORD

The whirlwind rise of sophisticated mobile devices and high-speed wireless networks is transforming the Internet and how our society interacts with it. Americans are rapidly embracing mobile technologies and the opportunities they create. In fact, adoption of mobile Internet is proceeding faster than earlier technologies, including the television (DeGusta, 2012). The use of mobile Internet applications skyrocketed between July 2011 and October 2012, rising by double-digit percentage points across generations. Mobile communications are now nearly ubiquitous—almost 9 out of 10 Americans ages 25 and older reported that they use mobile phones. The data suggest that the use of mobile devices for communications and information access has expanded exponentially and is now deeply ingrained in the American way of life.

Data from the Census Bureau's October 2012 Current Population Survey (CPS) show that 42 percent of mobile phone users ages 25 and older reported

* This is an edited, reformatted and augmented version of report issued October 2014.

browsing the Web on their devices, up from 33 percent in 2011. Similarly, email use on mobile phones grew from 33 percent to 43 percent of users, and use of social networking applications increased from 22 percent to 30 percent. Over half (54 percent) of mobile phone users reported using their devices to take photos or video, representing a 13 percentage-point increase in just 15 months.

Additionally, 88 percent of Americans over 25 reported in the CPS that they used mobile phones, and disparities in adoption of these devices shrunk in a number of key groups (including individuals lacking post-secondary education, those with annual family incomes below $25,000, and disabled Americans) more rapidly compared to the country as a whole. According to Pew Research Center, as of January 2014, more than half (55 percent) of all Americans 18 years or older now have smartphones, and 42 percent own tablet computers (Pew Research Center, 2014).

When combined with advances in mobile Internet connectivity, some form of broadband, whether fixed or mobile, is now available to almost 99 percent of the U.S. population. That said, while broadband adoption has grown from 4 percent to 72 percent of households since 2000, 28 percent of households still lacked home broadband connections by October 2012. Over a quarter of these non-users, representing over 7 percent of American households, did not go online at home primarily because it was too expensive. The continued persistence of financial and other barriers to Internet use is an urgent problem for policymakers because the "Internet has become integral to daily life in the United States" (NTIA & ESA, 2013). Recently, the President reiterated the importance of Internet access to American society:

> If you think about the history of America, we united a continent by rail. We stretched a network of highways from sea to shining sea. We brought light to dark and remote areas. We connected the world through the Internet, through our imagination. All these projects created jobs. All these projects grew our economy. They also unified the country and they are unifying the world.
>
> —President Obama on Technology in Schools, June 6, 2013

This report is the latest edition of our *Digital* Nation series, and it is based on data collected by the U.S. Census Bureau at the request of NTIA. Against this backdrop of data, today's Internet-connected society raises a range of significant policy issues, including questions about spectrum use, universal service, copyright, privacy, security, and the economy itself.

The data collected from NTIA's Broadband Technology Opportunities Program (BTOP) and the State Broadband Data and Development grants that supported data collection for the National Broadband Map, as well as the CPS data, provide insight for the President's wireless initiative, the Federal digital literacy programs, and the FCC's universal service reform initiative. These efforts will also help state and local policymakers throughout the country, in cooperation with private partners, to make sound, fact-based decisions that will improve America's productivity, ingenuity, world competitiveness, and standard of living.

Consistent with this Administration's resolve to maintain a transparent and open government, the survey data utilized in this report are publicly available for use and analysis by academic and policy researchers and others interested in the state of Internet use in America.

EXECUTIVE SUMMARY

Exploring the Digital Nation: Embracing the Mobile Internet builds on the last *Digital Nation* study (NTIA & ESA, 2013) and examines data from the 2012 Current Population Survey (CPS) describing Americans' use of computers and the Internet. Data from this survey depict the accelerating popularity of mobile phone-based Internet use across American demographic groups. The Census Bureau collected data from over 53,000 households in the October 2012 CPS Computer and Internet Use Supplement, forming a representative sample of the 122 million households across the United States. NTIA's analysis of this dataset, summarized below, reveals a mobile Internet explosion, accompanied by narrowing but persistent disparities in computer ownership and Internet use overall.

As in all installments of *Digital Nation*, this report includes NTIA's analysis of the demographic characteristics of home Internet users and the technologies they use to go online, as well as the alternative locations where they use the Internet. The report also presents data on the primary reasons why some Americans do not access the Internet from their homes. These data can inform further research and policy that strives to eliminate disparities by increasing the availability of affordable broadband in the United States and promoting broadband adoption by all Americans.

Highlights of this report include the following:

Online On-the-Go with Mobile Phones

Mobile Internet usage has grown substantially across demographic categories.

- In October 2012, mobile phone users were exploiting their devices' enhanced capabilities for much more than just voice communications. Over a 15-month period, the proportion of users who downloaded mobile applications jumped by ten percentage points, from 22 to 32 percent. In addition, 54 percent of mobile phone users at least 25 years of age reported taking photos or videos with their phones, 43 percent stated they checked or sent email, and 42 percent browsed the Web. Each of the latter two activities increased in prevalence by 9 percentage points since the July 2011 survey.
- Eighty-eight percent of Americans ages 25 and older used mobile phones by October 2012, up 2 percentage points from July 2011. This growth reflects an encouraging narrowing of the adoption gap among historically disadvantaged groups. For example, mobile phone use increased 4 percentage points each among individuals with family incomes below $25,000 (73 percent to 77 percent) and people with disabilities (68 percent to 72 percent).
- Use of Internet-dependent applications on mobile phones varied significantly based on educational attainment, family income, and population density. For example, among mobile phone users ages 25 and older, 57 percent of college graduates stated they checked or sent email, and 54 percent of users with college degrees browsed the Web on their mobile phones, compared to 19 percent and 21 percent, respectively, of those with no high school diploma. A 36 percentage-point gap exists between the 63 percent of mobile phone users with family incomes of $100,000 or more who used their devices for email and the 27 percent of users with family incomes below $25,000 who did so. Even as mobile phones approach ubiquity and traditional disparities in adoption shrink, there is evidence of a socioeconomic-based digital divide in usage patterns. Examination of population density also revealed differences in users' propensity to browse the Web on their mobile phones, with 44 percent of urban dwellers reporting they used their mobile phones for that purpose compared to 31 percent of rural residents.

- NTIA modeled the probability that mobile phone users ages 25 and older checked or sent email, browsed the Web, downloaded apps, and used online social networks through a technique known as multivariate regression analysis. The regression models enabled NTIA to study how various demographic and geographic characteristics correlate with these activities when holding other factors constant. The results were largely consistent across activities, with higher income and education levels strongly correlating with an increased probability of undertaking the activities.
- The regression models predict that female mobile phone users were 5 percentage points more likely to use social networks on their devices than their male counterparts. Further, the results suggest associations between geographic region and Internet-based mobile phone activities. The model forecasts that users in the West were between 6 and 8 percentage points more likely to use their phones for the studied purposes than their counterparts in the Northeast, and users in the South and Midwest exhibited smaller increases in these activities.

The Diminishing Digital Divide: Adoption Demographics

According to 2012 CPS data, U.S. broadband adoption continued to grow.

- Seventy-two percent of U.S. households used broadband at home in 2012, while home computer use grew to 79 percent, representing a 3 percentage-point gain in both activities from 2011 (69 percent and 76 percent, respectively). Two percent of households continued to rely on dial-up service, which raised the percentage of households engaged in any home Internet use to approximately 75 percent.
- Over the five years from 2007 to 2012, home broadband use by persons 65 and older rose dramatically, increasing from 32 percent to 47 percent, a 15 percentage-point increase.
- Persons 16 years and older who use the Internet at any location grew from 151 million in October 2007 to 187 million in October 2012, an increase of over 18 percent after adjusting for population growth.
- Libraries were important locations of Internet access across all income and educational brackets (used by 11 percent of households nationally), but especially so for unemployed householders (20

percent), households with school-age children (18 percent) and African Americans (16 percent).

- In 2012, 83 percent of households headed by an employed person reported using the Internet at home, compared to 70 percent of unemployed respondents. In 2007, only 71 percent of employed householders and 56 percent of unemployed householders used the Internet at home, resulting in a 13 and 15 percentage-point increase, respectively, over five years.

Persistent Gaps in Home Internet Use

In 2012, a significant portion—28 percent—of American households did not use broadband at home. Households commonly cited several primary reasons for non-use.

- **No Need or Interest:** Over time, U.S. households without the Internet at home have most often cited a lack of need or interest as the main reason why they did not go online at home. Although 48 percent of non-using households gave this reason in both 2011 and 2012, the figure rose from 39 percent in 2003. Consistently, White and Asian American households were most likely to express a lack of interest in going online at home. American Indian and Alaska Native households have been the least likely to say they lacked a need or interest in going online at home. In 2012, households led by people with disabilities citing lack of interest or need for home Internet use increased to 56 percent of non- users, compared to 51 percent in 2011. Households with school-age children that lacked home Internet service were 7 percentage points less likely to state they did not need to use the Internet at home at 47 percent, compared to 54 percent of those without school-age children. Non-using householders ages 65 or older were the least interested in going online at home, and the frequency with which seniors gave this response rose 3 percentage points from 67 percent in 2011 to 70 percent in 2012.
- **Expense:** Affordability concerns deterred the next largest segment of households from going online at home. In 2012, as in past years, this reason was the second most popular explanation for no home Internet use, at 29 percent of non-using households. Viewed from a different perspective, these figures indicate that 7 percent of *all* American

households do not have Internet service at home because of the cost. Among households not online, unemployed householders (58 percent), the youngest householders 15 to 24 years old (50 percent), those with family incomes below $25,000 (32 percent), and homes with children in school (30 percent) responded most often in 2012 that cost concerns prevented them from using the Internet at home. While White households not online (20 percent) continued to be the least concerned about the high cost of accessing the Internet in their homes, the proportion of other ethnic groups citing expense grew between 2011 and 2012. During that period, householders of American Indian or Alaska Native origin who did not go online at home became dramatically more likely to cite expense, increasing by 17 percentage points from 25 percent to 42 percent, followed by a 9 percentage-point increase among Asian American households. More urban households (30 percent) than rural residents (25 percent) that did not use the Internet at home stated that cost was the main reason for non-use.

- **No or Inadequate Computer:** This reason for not using the Internet at home has continued to decrease in prominence, declining by 2 percentage points between 2011 and 2012 to 11 percent. Data from the 2012 CPS reflected slight differences based on race and ethnicity, with non-using American Indian and Alaska Native households being the least likely at 8 percent to cite the lack of an adequate computer. This figure suggests a sharp 10 percentage-point decline among this group from 2011. Age, education, income, employment status, population density, gender, and the presence of school-age children at home only nominally differentiated households explaining they lacked an adequate computer as their main reason for not connecting to the Internet at home.

Introduction

There is considerable evidence that high-speed (broadband) Internet has a positive effect on a nation's economy and well-being (NTIA & ESA, 2013).[1] For example, about 752,000 U.S. jobs come from the mobile applications ("apps") industry, which followed the rapid spread of smartphones (Mandel, 2013). One analyst estimates that each American consumer can save $8,674 annually by using broadband in a variety of ways (Internet Innovation

Alliance, 2013). For example, a recent study found that 69 percent of surveyed baby boomers and seniors over age 54 used their smartphones while shopping in local stores, and half of them did so primarily to comparison shop and find discounts (Cisero, 2014). By the end of 2014, U.S. consumers will spend an estimated $114 billion in online retail sales on smartphones and tablets, which will account for 29 percent of all e-commerce sales, according to analysts' projections (Mulpuru, 2014). As NTIA stated in its June 2014 progress report on freeing up spectrum for commercial use, "America's future competitiveness and global technology leadership depend on access to radio spectrum—the lifeblood of smartphones, tablets, and other data-hungry wireless devices."

Deployment of mobile wireless broadband networks (with speeds of at least 3 Mbps download and 768 Mbps upload) has proliferated in the United States in recent years, resulting in availability of service to 97.5 percent of the nation's population as of June 30, 2013. Wired broadband availability to Americans by individual technology—fiber (25.2 percent), DSL (74.0 percent), and cable (87.5 percent)—trails the coverage of mobile wireless networks (NTIA & FCC, 2014). There is good news concerning higher- speed Internet as well. Ninety percent of Americans had access to 4G wireless broadband, defined as service with download speeds of at least 6 Mbps, as of the end of 2012. The increase over June 2010 is impressive, gaining approximately 64 percentage points during the span (NTIA, 2013).

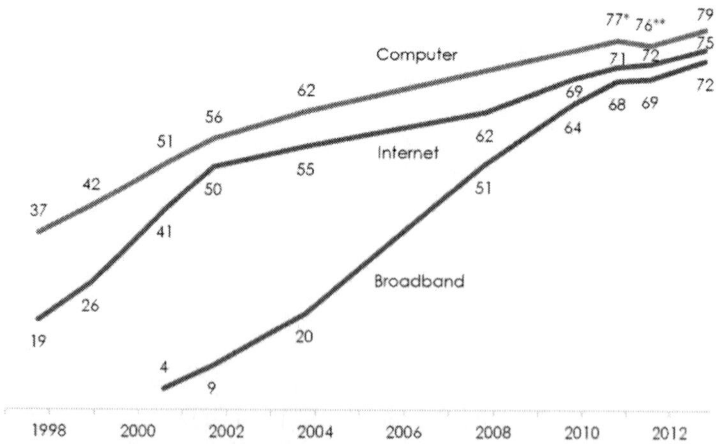

* Includes handheld devices such as smartphones and tablets (2010 only).
** Includes tablets but not smartphones (2011-2012).

Figure 1. Overview of Household Adoption Rates by Technology, Percent of U.S. Households, 1997-2012.

Broadband adoption and computer ownership by U.S. households continued their rise in 2012.[2] Research suggests that adoption of new technologies typically resembles an S- shaped curve over time (Rogers, 1995). The slopes in Figure 1 have decreased over time but there is little evidence that any of the technologies depicted have reached saturation. Factors that affect the shape of the curve include consumer awareness of the technology, its affordability, accessibility, and ease of use (NTIA & ESA, 2000).

The use of mobile devices continued to increase in 2012. Eighty-eight percent of Americans ages 25 and older reported using mobile phones in October 2012. Once online, mobile phone users increasingly used their devices to send and receive email, browse the Web, access social networks, and utilize other applications that offer increased productivity or entertainment in their busy lives.

Smartphones and their diverse applications have led the mobility movement. Popular with the general public, smartphones are also important to people with disabilities for reasons beyond mere convenience. Smartphone applications can reveal one's location via GPS or provide directions to a desired destination. To help people with vision disabilities, researchers in Pakistan have developed a smartphone app that monitors a user's location and distance walked from a destination to warn of imminent nightfall. The application is geo-aware and so knows the time of sunset around the world, and with its access to online mapping software, the app can offer the potentially vulnerable user shortcuts for the return trip (Ahsan, Khan, & Salam, 2013).

Smartphones can also aid the unemployed. Some 77 percent of job seekers have already begun to use smartphone apps to give them an advantage in job-seeking. Application developers are creating tools that enhance one's ability to learn about and secure a new job and to customize and export a résumé expeditiously. Another uses geo-targeting to find industry-specific contacts that are open to career networking. One app permits busy job-seekers to create short videos with answers to employer questions (Jackson, 2012).

The sections that follow illustrate the results of the Current Population Survey (CPS) Computer and Internet Use Supplement, collected by the Census Bureau in October 2012 at the request of NTIA. In the next section, we provide an analysis of mobility in the online world. The data suggest that the use of mobile devices for communications and information access has expanded exponentially, becoming an essential element of American daily life. The focus then turns to adoption, with a longitudinal analysis of devices of choice, Internet technologies, demographics, geography, and locations of

Internet use. We believe that, while progress continues in home broadband adoption, disparities among groups and areas persist, and libraries and other public access points provide alternative venues for Internet use. The final section contains new evidence as to why some U.S. households remained unconnected in 2012, providing key insights for policymakers seeking to expand Internet usage to all Americans.

THE MOBILE TRANSFORMATION

Digital Nation reports have historically focused on trends in computer and Internet use in the United States that highlight policy choices that might further accelerate adoption. Getting more Americans online remains vital to the nation's future. However, policymakers should note that in recent years there have been fundamental changes in the way people use the Internet. Through a combination of faster wireless data networks, advances in the miniaturization of computing hardware, and innovations in human-computer interaction, the Internet is now in people's pockets and at their fingertips, available to many at almost any time and location.

To date, adoption of mobile computing technologies has primarily been in the form of mobile phones with advanced capabilities ("smartphones") and tablet computers. NTIA tracked the use of mobile phones in the July 2011 and October 2012 CPS Supplements, and the results are discussed in this section. The 2012 survey, which was less detailed than the 2011 survey, did not distinguish usage of tablets from desktop and laptop computers. Data from upcoming CPS Supplements will enable us to track growth in tablet adoption since the 2011 survey, which showed that 9 percent of Internet users reported using tablets to go online (NTIA and ESA, 2013). Research by the Pew Research Center suggests that 42 percent of Americans ages 18 and older owned tablet computers by January 2014; that number rises to 50 percent if e-readers are included (Pew Research Center, 2014).

Eighty-eight percent of Americans ages 25 and older used mobile phones as of October 2012. While this represents a small (2 percentage-point) increase from July 2011, mobile phone usage is evolving in two important ways. First, mobile phones are becoming more common among historically disadvantaged groups; the adoption gap is shrinking across demographic and socioeconomic groups. Second, Americans are taking advantage of the wide range of new functionalities made possible by the fast Internet connections and advanced hardware present in contemporary mobile phones. However, there are some significant disparities in phone usage patterns. Among mobile phone users, use of Internet-based applications is primarily a function of income, education, and age, even after controlling for a range of demographic variables.

Mobile Phone Adoption

The July 2011 CPS Supplement revealed different levels of mobile phone adoption based on income, education, race, and other factors, suggesting this technology is subject to a digital divide similar to those found in computer and Internet use (as described in previous *Digital Nation* reports). We are encouraged, however, that disparities in mobile phone use dropped significantly by the time of the October 2012 data collection. The small overall increase in mobile phone adoption during this time masks a more interesting story: adoption among historically lagging groups increased more quickly than among groups where mobile phone use was already extremely common. Among adults ages 25 and older, 96 percent of those with annual family incomes of $100,000 or greater used mobile phones in 2012, a nominal increase from 95 percent in 2011. During the same time period, however, those with family incomes below $25,000 became 4 percentage points more likely to use mobile phones, as adoption in this group increased from 73 to 77 percent (see Figure 2).

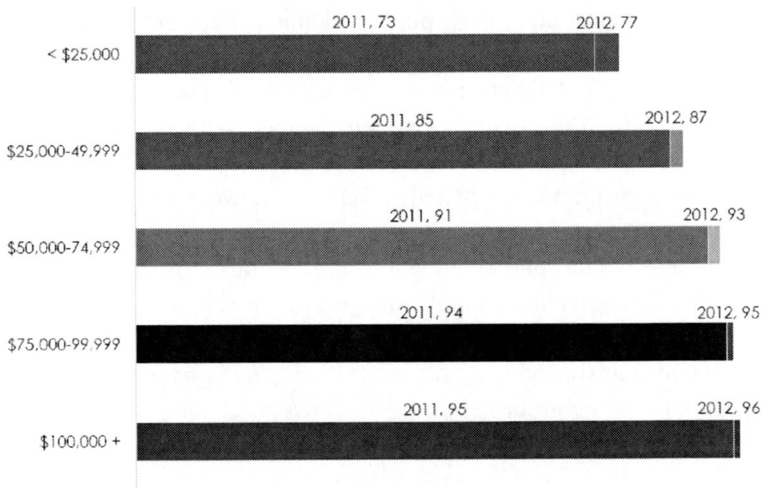

Figure 2. Mobile Phone Use by Family Income, Percent of Persons Age 25+, 2011-2012.

There was also accelerated growth in mobile phone use among other historically lagging groups. Adoption among Americans with disabilities grew at twice the rate of their non- disabled counterparts. Mobile phone use increased from 68 percent of those who identified themselves as having a disability in 2011 to 72 percent in 2012, while adoption among those who did not self-identify as having a disability grew from 88 to 90 percent. Similarly, individuals with lower education levels narrowed the adoption gap with their highly-educated counterparts. Seventy-two percent of Americans who lacked a high school diploma or equivalent used mobile phones by 2012, an increase from 68 percent in 2011, and adoption among those who graduated from high school but did not continue their education increased from 79 to 83 percent. In contrast with the 4 percentage-point increases among those groups, mobile phone use among college graduates increased by just one percentage point, from 94 to 95 percent.

Mobile phone use among rural[3] Americans also increased significantly between 2011 and 2012. Eighty-five percent of rural dwellers reported using a mobile phone in 2012, a 5 percentage-point increase from the 80 percent reported 15 months earlier. In contrast, the adoption rate among urban Americans only increased by 2 percentage points, from 86 to 88 percent (see Figure 3). The narrowing of this adoption gap based on population density is particularly promising because rural areas have historically experienced less-extensive build-out of network infrastructure than urban areas (NTIA & FCC, 2014).

Racial disparities in mobile phone adoption were small in 2011 and appeared to nearly vanish in 2012. In 2011, 86 percent of Whites used mobile phones, compared to 84 percent of African Americans and 83 percent of Hispanics. In 2012, however, adoption among African Americans and Hispanics grew to 87 percent each, while adoption among Whites grew more slowly to 88 percent; the estimated difference between those groups is no longer statistically significant (see Figure 4).

Trends in mobile phone adoption during this 15-month period are encouraging. Demographic disparities appear to be shrinking, and in at least one case may have disappeared—though it is important to recognize that substantial gaps remain and will only close if adoption continues to grow more quickly among lagging groups.

Figure 3. Mobile Phone Use by Population Density, Percent of Persons Age 25+, 2011-2012.

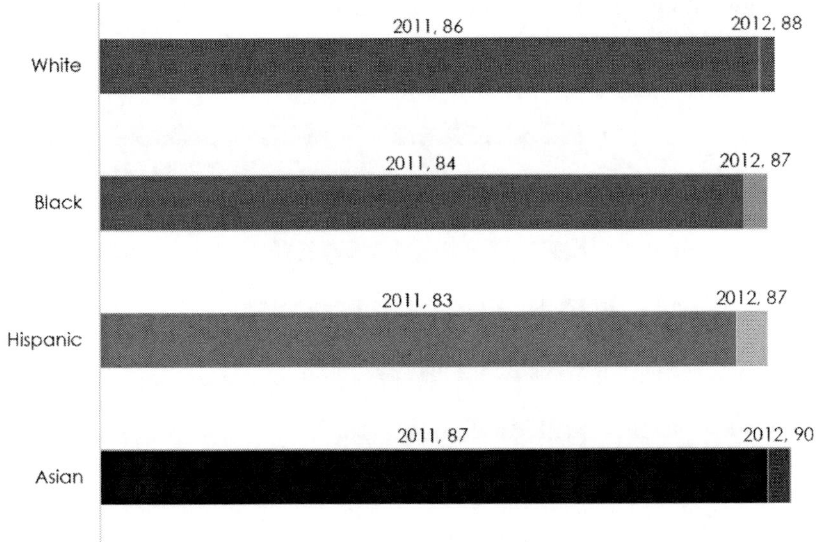

Figure 4. Mobile Phone Use by Race, Percent of Persons Age 25+, 2011-2012.

Use of Mobile Applications

Contemporary mobile phones are not mere telephones. In the rapid transition to advanced devices on high-speed data networks, mobile phone users are engaging in a dramatically expanded range of activities. Beyond phone calls and text messaging, mobile phones offer constant access to Internet applications like email and Web browsing, location-based services, and the ability to take pictures and listen to music.

Use of these applications is skyrocketing. Usage of non-voice applications grew rapidly between July 2011 and October 2012, with application usage growing 10 percentage points or more over the 15-month period in many cases. For example, over half (54 percent) of American mobile phone users ages 25 and older used their devices to take photos or video by 2012, a 13 percentage-point increase. The use of mobile phones for checking or sending email grew from 33 percent of users to 43 percent, and Web browsing increased from 33 to 42 percent. The proportion of mobile phone users utilizing maps and other GPS-based applications grew by 10 percentage points to 34 percent in 2012. Other uses of Internet-connected mobile phones, such as social networking and downloading apps, also saw considerable increases in popularity (see Figure 5).

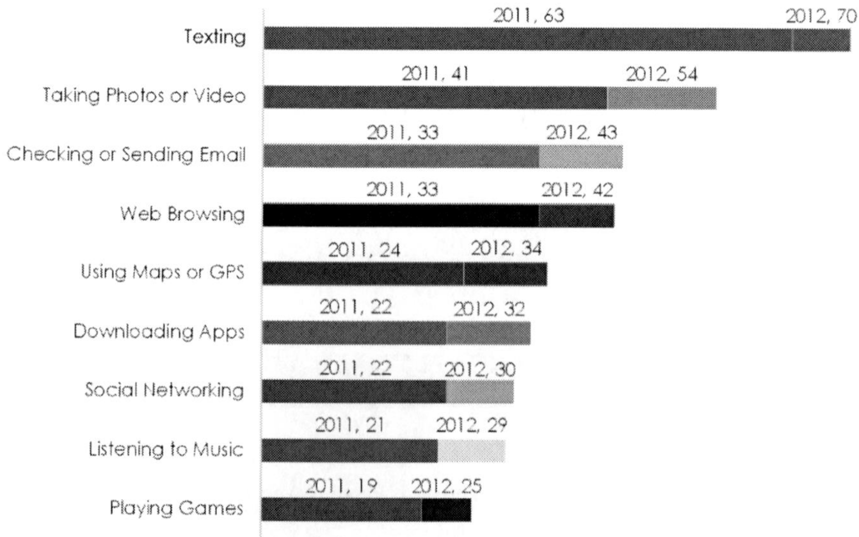

Figure 5. Activities Americans Conduct on Mobile Phones, Percent of Mobile Phone Users Age 25+, 2011-2012.

NTIA focused on four reported Internet-dependent activities: checking or sending email, Web browsing, downloading apps, and social networking to determine if there are disparities in usage patterns among different groups of mobile phone users. Data show that there are dramatic differences in mobile phone usage based on educational attainment. Among mobile phone users ages 25 and older, 57 percent of college graduates reported checking or sending email on their devices in 2012, compared to 45 percent of those with some college credit, 29 percent of high school graduates, and 19 percent of those with no diploma. Similarly, 54 percent of users with college degrees browsed the Web on their mobile phones, compared with just 21 percent of those without high school diplomas. There were also significant disparities in downloading apps and social networking based on education (see Figure 6).

Use of mobile phone-based Internet also differs greatly based on income. Sixty-three percent of mobile phone users with family incomes of at least $100,000 checked or sent email with their devices, compared with 52 percent in the $75,000-99,999 range, 45 percent in the $50,000-74,999 range, 36 percent in the $25,000-49,999 range, and 27 percent with family incomes below $25,000. Web browsing, downloading apps, and social networking were similarly stratified.

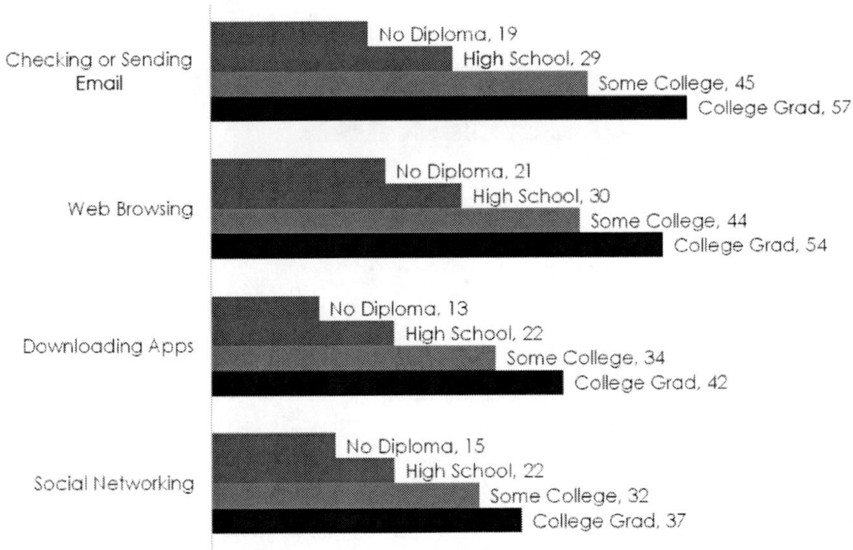

Figure 6. Activities Americans Conduct on Mobile Phones by Education, Percent of Mobile Phone Users Age 25+, 2012.

In stark contrast to the modest and shrinking mobile phone adoption gap between urban and rural Americans, use of Internet-based applications on those mobile phones varied dramatically by population density. For example, while 45 percent of urban mobile phone users checked or sent email with their devices, only 29 percent of their rural counterparts reported doing so. Additionally, rural users were 13 percentage points less likely to browse the Web on their phones, 12 percentage points less likely to download apps, and 8 percentage points less likely to use social networks (see Figure 7). Slower wireless network speeds in rural areas, in addition to demographic differences between urban and rural dwellers, may partially explain these disparities (NTIA & FCC, 2014).

Similar to mobile phone adoption generally, Internet use on devices does not vary dramatically by race. The one exception is that Asian American mobile phone users were significantly more likely to check or send email, browse the Web, download apps, and use online social networks with their devices. For example, 51 percent of Asian American users reported checking or sending email on their mobile phones, compared with 43 percent of Whites, 40 percent of African Americans, and 42 percent of Hispanics (see Figure 8). These differences may be due in part to other demographic characteristics, like income and education levels.

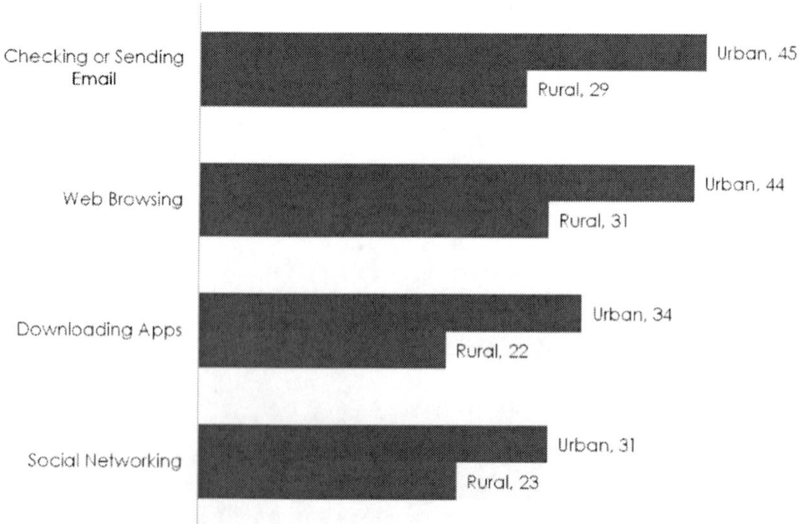

Figure 7. Activities Americans Conduct on Mobile Phones by Population Density, Percent of Mobile Phone Users Age 25+, 2012.

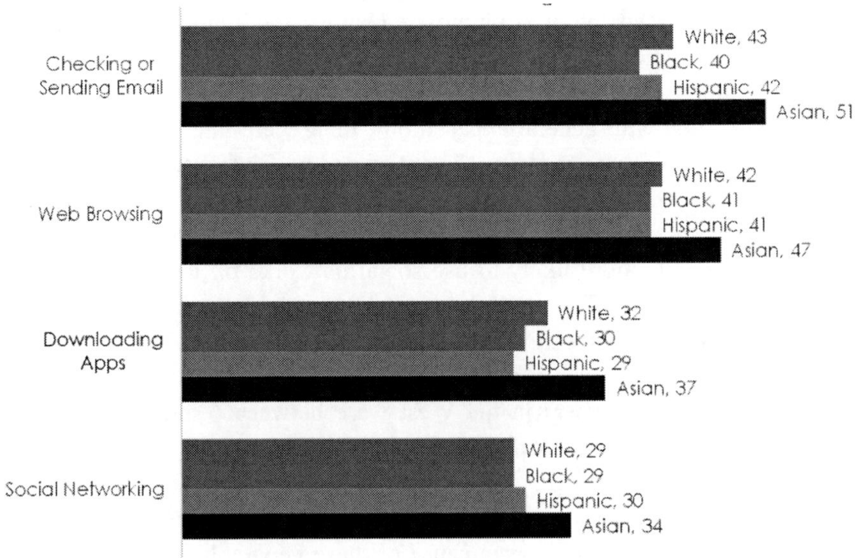

Figure 8. Activities Americans Conduct on Mobile Phones by Race, Percent of Mobile Phone Users Age 25+, 2012.

To better understand why certain factors are associated with different patterns in mobile application use, NTIA used multivariate regression analysis, a statistical technique used to predict the marginal effects of different characteristics by holding constant other known factors. NTIA performed four regression analyses, modeling the probability that mobile phone users ages 25 and older check or send email, browse the Web, download apps, and use online social networks. In each case, income, education, age, race, gender, disability, citizenship status, employment, presence of school-age children in the household, household size, population density, and geographic region were held constant. The results were largely consistent—there was a strong correlation between higher income and education levels and increased probability of undertaking the various phone-based activities. For example, when holding other factors constant, mobile phone users with family incomes of $100,000 or more were 19 percentage points more likely to browse the Web on their devices than their counterparts in the below $25,000 category. Similarly, users with college degrees were 22 percentage points more likely to use email on their phones than those without college diplomas, after controlling for other variables. Being employed was associated with higher rates of Internet-based activities on mobile phones, and living in an urban area correlated with higher usage rates relative to rural dwellers.

While the results for income, education, employment, and population density are in line with general expectations, there were interesting results by gender and geographic region. There was no statistically significant relationship between gender and using email, browsing the Web, or downloading apps, but the model predicts that female mobile phone users were 5 percentage points more likely to use social networks on their devices than their male counterparts. This suggests that gender may play a role in the choice to engage in social networking from a mobile phone. Additionally, focusing on associations between geographic region and Internet-based mobile phone activities revealed that users in the West were between 6 and 8 percentage points more likely to use their phones for the studied purposes relative to their counterparts in the Northeast, and users in the South and Midwest exhibited smaller (but in most cases, still statistically significant) increases in these activities (see Figure 9). The exact cause of these geographic disparities after controlling for income, education, and other factors is worth further examination.

Equally interesting are the correlations not exhibited in these regression analyses, most notably race and ethnicity. Consistent with the modest observed racial differences in mobile phone activities, the marginal effects of race on these phone-based Internet uses were small; in most cases the differences were not statistically significant. The model suggests that income and education are the largest predictors of mobile phone use habits, with employment status, geography, and a few other factors also playing roles. No one study of survey data can establish that any of these variables actually causes certain usage patterns. This analysis can only demonstrate that the factors appear to be correlated after controlling for other variables that are also thought to contribute.

Examination of the October 2012 CPS Supplement data on mobile phone use suggests a need for further research and policy focus on *how* people are using their devices, rather than mere ownership. Simply having such a device does not reliably indicate that a user is taking advantage of the advanced applications made possible by recent innovations in mobile technologies. As mobile phones approach ubiquity and traditional disparities in adoption shrink, there is evidence of a new socioeconomic-based digital divide in usage patterns. Thus, policymakers should focus on addressing its possible causes, including the additional cost of mobile data plans, limited availability of high-speed networks in rural areas, or gaps in digital literacy. This is an essential step in extending to all Americans the opportunities created by carrying the entire Internet in their pockets.

GETTING ONLINE

Although gaps remain, broadband adoption continues to grow across demographic and geographic groups. Data also suggest some interesting variations in broadband adoption based on a number of factors. For example, households with school-age children report higher rates of adoption, possibly indicating greater demand among younger individuals or perhaps students' need to use the Internet for homework and studying.

	Using Email	Web Browsing	Downloading Apps	Social Networking
Family Income, Relative to Less Than$25,000				
$25,000-49,999	4	3	4	4
$50,000-74,999	9	7	7	7
$75,000-99,999	13	11	12	8
$100,000 or More	22	19	18	12
Education, Relative to No High School Diploma				
High School Diploma	6	5	4	4
Some College	16	14	11	9
College Degree	22	18	15	12
Employment Status, Relative to Employed				
Unemployed	-6	-4	-4	-5
Not in Labor Force	-6	-5	-4	-3
Region, Relative to Northeast				
Midwest	N/A	3	3	3
South	5	5	4	5
West	7	8	8	6
Other Factors				
Female, Relative to Male	N/A	N/A	N/A	5
Urban, Relative to Rural	8	6	6	3

Note: All listed marginal effects are statistically significant at the 95 percent confidence level. See Appendix B, Tables 1-4 for complete regression results.

Figure 9. Marginal Effects of Demographics on Selected Mobile Phone Activities, Percentage-Point Change for Mobile Phone Users Age 25+, 2012.

Internet at Home: Devices, Technologies, and Trends

Computer Use at Home

Data collected in 2012 on Americans' digital device choices show that 79 percent of households reported having computers at home, a 3 percentage-point increase from 2011. This also represents a 28 percentage-point increase in home computer use since 2000.

EXPANDING BROADBAND USE IN ARKANSAS THROUGH EDUCATION

When Connect Arkansas received a $3.7 million BTOP award to increase broadband adoption, the state ranked 49th in Internet subscribership and 47th in per capita income, according to FCC data. NTIA's data suggest that the state has significantly improved its broadband adoption rate during the last few years (see Figures 11 and 12). To encourage broadband adoption by Arkansans, the project offers digital literacy and computer basics training for adults and free refurbished computers to qualifying low income families. In addition to selling discounted reconditioned computers to the public, Connect Arkansas trains residents in business plan writing, the principles of entrepreneurship, and small business technology use. See the project website at http://www.connect-arkansas.org/home for more details.

In 2000, 61 percent of 25- to 44-year-old householders[4] reported having a computer at home, while only 24 percent of those 65 and older reported the same. In 2012, this number jumped to 85 percent for 25- to 44-year-olds and to 64 percent for those 65 and older, a 40 percentage-point increase among seniors. In 2001, only 54 percent of unemployed householders owned a computer, whereas by October 2012, 75 percent of households facing unemployment had this important resource at home to aid in job seeking.

In 2012, low-income households continued to lag behind wealthier households in having a computer at home. Only 57 percent of households earning less than $25,000 a year had a computer at home, while they were nearly ubiquitous in households earning $100,000 or more (97 percent). Householders with no high school diploma made strides in closing the computer-at-home gap by 31 percentage points over the last 12 years (2000:

18 percent to 2012: 49 percent); yet in 2012, they still lagged far behind those with a college degree or higher (94 percent).

Throughout the last decade, Asian American householders consistently reported the highest rates of home computer use, while the lowest percentage shifted over the years between African American, Hispanic, and American Indian and Alaska Native households. In 2012, African American householders reported the lowest prevalence of computers at home at 67 percent.

The gap is closing between computers at home in rural versus urban communities. In 2000, urban households were 11 percentage points more likely to have computers at home than their rural counterparts. By 2012, however, that gap narrowed to 8 percentage points.

Internet Use at Home

Internet use at home is continuing to rise, although not at the same rates observed during the earlier years of the 21st century. Between August 2000 and September 2001, the proportion of households going online at home jumped by 9 percentage points, from 41 percent to 50 percent. In 2012, 75 percent of households reported having Internet at home, representing a 3 percentage-point increase from 2011. By 2012, just 2 percent of households online at home relied solely on dial-up Internet service.

Different demographic and geographic groups reported disparate levels of home Internet use. For instance, in 2012 82 percent of 25- to 44-year-old householders had Internet at home, similar to 79 percent of 45- to 64-year-olds. Those 65 and over continued to lag behind other age brackets, with only 57 percent reporting home Internet use. Income levels also impacted Internet adoption at home. Low-income households were far behind their wealthier counterparts: Forty-nine percent of households making less than $25,000 used the Internet at home, compared to 96 percent of households making $100,000 or more.

In 2012, 40 percent of householders without a high school diploma reported having the Internet at home, a 28 percentage-point gain over the last 12 years, but significantly behind the 92 percent of home users with at least a college degree. Although 83 percent of employed householders used the Internet at home in 2012, such use has continued to rise among the unemployed as well, with 70 percent of unemployed householders having the Internet at home, compared to 50 percent in 2003. Importantly, households led by females continued to lag in home Internet use over the last 12 years. Although female heads of household narrowed the gap from a 10 percentage point disparity (2000, 2001), they continued trailing male heads of household by 6 percentage points in 2012 (72 percent adoption for female householders).

Similar to home computer use, while home Internet use increased during the last decade, adoption continued to be lowest among African American, Hispanic, and American Indian and Alaska Native householders. One of the largest increases occurred among Hispanic householders, with 64 percent reporting Internet at home in 2012, up from 24 percent in 2000. As with computer use at home, Asian American householders continued to lead in home Internet adoption, with 85 percent reporting having the Internet at home.

	Computer Use	**Internet Use**	**Broadband Use****
All Households	79	75	73
Family Income			
Less Than $25,000	57	49	48
$25,000-49,999	77	72	70
$50,000-74,999	90	88	86
$75,000-99,999	95	93	92
$100,000 or More	97	96	95
Education*			
No Diploma	49	40	39
High School Diploma	69	65	63
Some College	84	80	79
College Degree	94	92	91
Race and Ethnicity*			
White	82	79	77
African American	67	62	61
Hispanic	69	64	63
Asian American	87	85	84
American Indian or Alaska Native	68	58	56
Population Density			
Urban	80	76	75
Rural	72	66	63

* These are attributes of the householder.

** For the purpose of this analysis, households are considered to use broadband if they report home Internet use and list any type of connection other than dial-up, including "some other service."

Figure 10. Home Computer, Internet, and Broadband Use by Demographics and Population Density, Percent of Households, 2012.

In 2012, rural households made progress in closing the gap with urban households, with 66 percent online; yet they remained 10 percentage points behind their urban counterparts, who reported 76 percent penetration. Disabled householders were 27 percentage points less likely to have the Internet at home than their non-disabled counterparts at 52 percent and 79 percent, respectively.

Home Internet Use by State

Similar to previous *Digital Nation* reports, states experienced significantly different Internet adoption levels in 2012. For example, households in some western and northeastern states reported home Internet adoption rates significantly above the national average of 75 percent, including Oregon at 85 percent and New Hampshire at 83 percent. In contrast, certain states in southern regions and Appalachia reported the lowest adoption rates nationally, including Mississippi at 58 percent of households, and Louisiana at 61 percent (see Figure 11).[5]

The estimated growth in Internet use during this three-year period varied significantly among the states from 2009 to 2012. Some historically lagging states appeared to make considerable progress towards closing the gap with their peers. For example, while only 56 percent of households in Arkansas were going online at home in 2009, the adoption rate there grew rapidly through 2012, when 69 percent of households used the Internet at home. While Arkansas still lagged behind the national home Internet use rate of 75 percent in 2012, the state may have significantly improved its standing (see Figure 12).[6]

Technologies Used to Go Online at Home

In 2012, cable and DSL were the most popular technologies households used to go online, with 45 percent of online households using solely cable modems and 23 percent reporting DSL connections. Mobile broadband ranked a distant third, with 9 percent of online households using the technology exclusively. Although cable modems were consistently most popular across all age groups (between 44 percent and 47 percent of online householders in each category), DSL was more popular among householders 65 and over, at 27 percent, than among other age groups. Online householders younger than 25 were about as likely to choose mobile broadband alone (17 percent) as DSL alone (18 percent). Overall, DSL usage has decreased over time. While 33 percent of online households exclusively used DSL at home in 2010, that number dipped to 23 percent by 2012. Only 5 percent of households with home Internet reported using both cable and mobile broadband at home, which represents a 3 percentage-point increase from 2010. Dial-up continued its decline as a choice for Internet service, standing at only 2 percent in the 2012 data collection. Dial-up service has always been most common among householders 65 and older, dropping, however, from 92 percent in 2000 to 4 percent in 2012.

A household's choice of broadband technology did not vary dramatically based on employment status, gender, disability status, or the presence of school-age children in the household. Across each of these categories, the overall averages remained consistent. Interestingly, 13 percent of online households with family incomes below $25,000 used mobile broadband exclusively. The exclusive use of mobile broadband at home was also most common among online heads of household lacking a diploma (12 percent), compared to the other education categories (7 percent for those with a college degree or higher). Four percent of those without a high school diploma reported utilizing dial-up alone, while only 1 percent of users with a college degree or higher did so. Seven percent of heads of household online at home with a college degree or higher accessed the Internet through fiber, while only 3 percent of those without a high school diploma reported the same. Six percent of online householders with college degrees reported use of both cable and mobile broadband at home, while only 3 percent without a high school diploma reported the same.

When considering employment status, mobile broadband use to access the Internet at home has grown most among the unemployed, rising from 8 percent in 2010 to 13 percent in 2012. After remaining flat between 2010 and 2011, growth in the use of both cable and mobile by households across income brackets occurred in 2012, while growing most significantly to 8 percent among households with incomes of $100,000 or more.

In 2012, cable modems were least prevalent among online American Indian and Alaska Native householders (39 percent), compared with their counterparts from other ethnic groups. DSL was the least popular among Asian American householders (20 percent). Mobile broadband adoption was highest among American Indians and Alaska Natives (14 percent), African American (13 percent), and Hispanic (12 percent) householders, and least so among Asian Americans (5 percent). American Indians and Alaska Natives reported some of the highest rates of home mobile broadband use and lowest rates of cable modem use; this may be partly due to a lack of cable broadband availability in regions with a high concentration of these households.

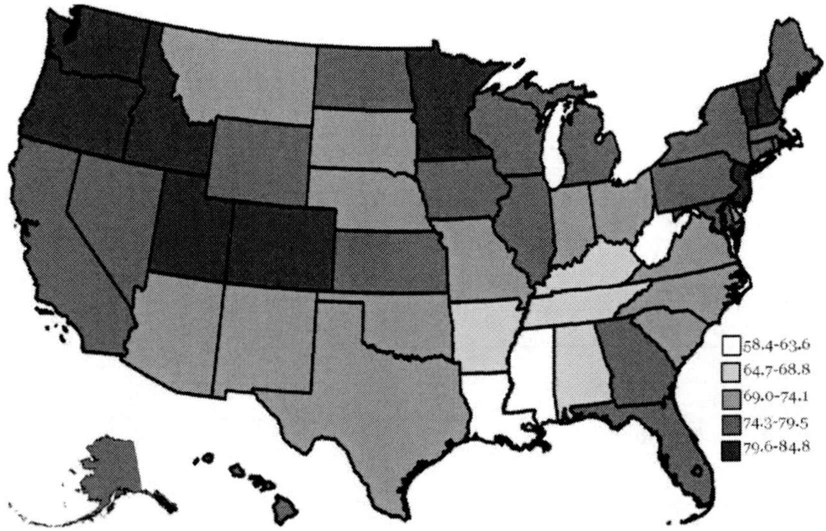

Note: See Appendix B, Table 5 for complete results.

Figure 11. Internet Use at Home by State, Percent of Households, 2012.

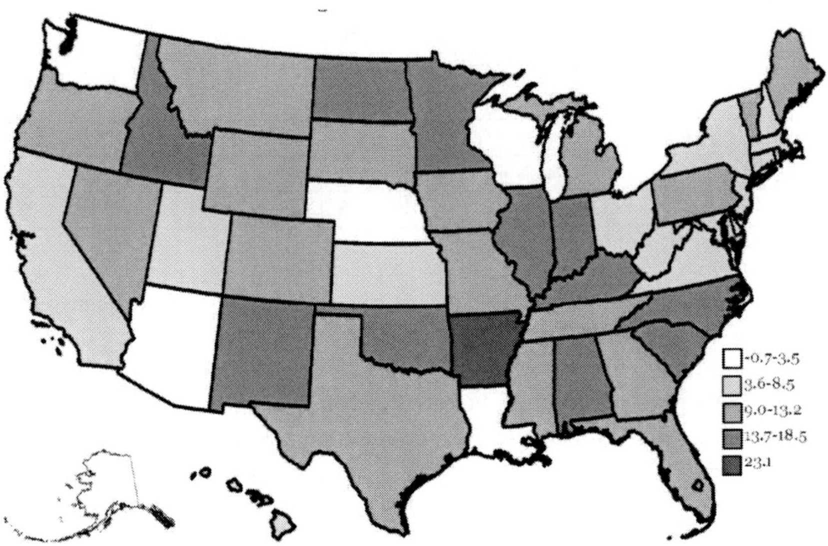

Note: See Appendix B, Table 5 for complete results.

Figure 12. Estimated Growth of Internet Use at Home Over Time by State, Percentage of Households, 2009-2012.

	Cable Only	DSL Only	Fiber-optic Only	Mobile BB Only	Cable and Mobile
All Online Households	45	23	5	9	5
Family Income					
Less Than $25,000	43	25	4	13	3
$25,000-49,999	44	25	4	10	4
$50,000-74,999	46	24	6	8	5
$75,000-99,999	45	24	6	7	6
$100,000 or More	45	20	8	6	8
Education*					
No Diploma	43	25	3	12	3
High School Diploma	43	25	5	10	3
Some College	44	24	5	10	5
College Degree	46	22	7	7	6
Race and Ethnicity*					
White	45	24	6	8	5
African American	42	24	5	13	5
Hispanic	41	24	5	12	4
Asian American	53	20	7	5	5
American Indian or Alaska Native	39	26	2	14	5
Population Density					
Urban	47	21	6	9	5
Rural	32	35	2	11	3

* These are attributes of the householder.

Figure 13. Selected Technologies Used to Go Online at Home by Demographics and Population Density, Percent of Online Households, 2012.

Four percent of online rural households relied on dial-up Internet service in 2012, while only 1 percent of their urban counterparts used dial-up service exclusively at home. Thirty-five percent of rural households with home Internet service used DSL, while only 21 percent of online urban households did the same. Forty-seven percent of urban users utilized cable alone, while only 32 percent of rural dwellers reported accessing the Internet this way. Satellite use was higher in rural areas at 5 percent, versus 2 percent reported by online urban households. Mobile broadband use was not significantly differentiated by geography: 11 percent of rural Internet-using households relied on this technology, while 9 percent of their urban counterparts reported the same. Only 2 percent of online rural households reported using fiber-optic connections for home Internet service, while 6 percent in urban areas reported utilizing this option. Of note, in 2011 and 2012 cable modem usage reported in rural areas was 15 percentage points lower than urban areas. However, during this same 15-month period, mobile broadband experienced a significant increase in rural communities, growing from 7 percent to 11 percent.

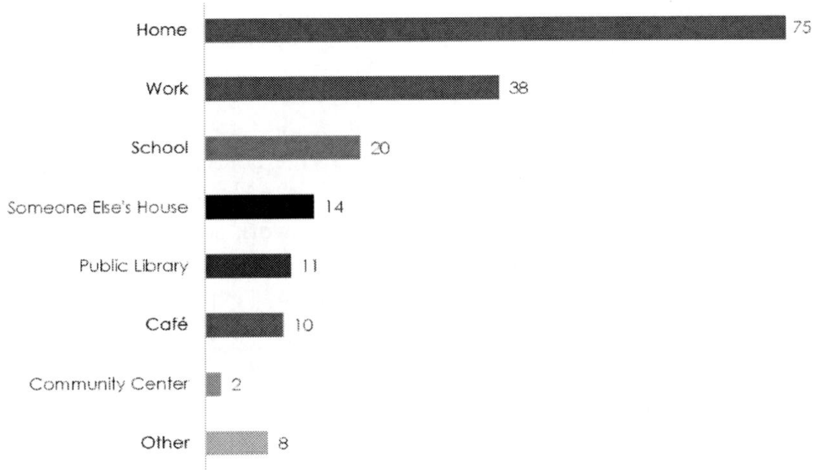

Figure 14. Internet Use Locations, Percent of Households, 2012.

Going Online in 2012: Connecting beyond the Home

Internet Use Outside the Home

In 2012, 42 percent of Americans used the Internet outside of the home, regardless of home use. Fifty-eight percent of 15- to 24-year-olds reported Internet use outside of the home, making them most likely to go online outside the home; only 16 percent of those ages 65 and older reported the same. Employed individuals were more likely than their unemployed counterparts to report going online outside the home, and persons with higher family incomes were more likely than low-income individuals to use the Internet outside the home. Internet use outside of the home was also higher among persons without a high school diploma (30 percent) than those whose education ended with a high school diploma (27 percent), which may be explained in part by Internet use at school. Hispanics reported the least amount of Internet usage outside of the home (33 percent), compared to 44 percent of both Whites and Asian

Americans. The biggest disparity was based on disability status, with only 19 percent of disabled Americans reporting Internet use outside of the home, compared to 47 percent of their non- disabled counterparts. Urban dwellers reported Internet use outside of the home 7 percentage points more often (43 percent) than their rural counterparts (36 percent). Of interest, whether school-age children lived in the home seems to have a notable impact on usage outside of the home, potentially pointing to school-age children's use of the Internet at school or libraries (no school-age children: 40 percent; school-age children: 45 percent).

Internet Use Locations Outside the Home

The 2012 survey asked respondents where the household's Internet users go online. Among households with a member who used the Internet at locations other than home, 38 percent identified work as a source of Internet access. Householders ages 25-44 reported that someone in the household used the Internet at work more often than other age groups (51 percent). While only 13 percent of households making less than $25,000 had someone going online at work, 70 percent of households making $100,000 or above did. Only 25 percent of householders with a high school diploma reported Internet use at work, while 59 percent of those with a college degree or higher did so. Significantly, survey results largely show a great deal of parity in adoption numbers between the genders, but Internet use at work seems to be an exception—40 percent of male householders reported someone going online at work, while 36 percent of females reported the same. Only 29 percent of Hispanic householders and 31 percent of African Americans reported Internet use at work, while 41 percent of Whites and 48 percent of Asian American householders reported the same. Only 13 percent of householders with a disability stated that someone accessed the Internet at work, compared to 42 percent of non-disabled householders. Among urban dwellers, 40 percent answered that someone in the household went online at work, while only 30 percent of rural households reported the same. While 47 percent of households with school-age children at home indicated that someone goes online at work, only 35 percent of households without school-age children reported the same.

Twenty percent of all households reported that someone in the household used the Internet at school. The authors observed a large gap between the lowest and highest income households: Only 15 percent of households making $25,000 and under stated that a member used the Internet at school, while 30 percent of households earning $100,000 and above responded similarly. Of interest, only 18 percent of White householders reported that someone within the household goes online at school, compared to 22 percent of African American householders and 26 percent each for of Hispanic and Asian American householders.

One possible explanation for this is that households with home Internet connections may rely less often on school Internet use (e.g., Whites at 79 percent and African Americans at 62 percent). This also raises questions, however, about why Asian American households, with an adoption rate comparable to White households in 2012, were also those most likely, along with Hispanic households to go online at school. These findings may benefit from further research and investigation. Only 16 percent of rural dwellers reported that someone in the household goes online at school, compared to 21 percent of households in urban areas. Eleven percent of households reported that someone went online at a public library. Use of public libraries to go online varies by age group: 15- to 24-year-old householders showed the highest utilization (16 percent), while householders 65 and older reported the lowest (5 percent). Of note, unemployed householders (20 percent) reported Internet use at public libraries much more often than their employed counterparts (11 percent). African American (16 percent), Asian American (14 percent), and Hispanic (12 percent) householders reported the most Internet use at public libraries, compared to 9 percent of White householders. Additionally, 18 percent of households with school-age children at home reported that someone went online at a public library, compared to only 8 percent of those without children at home. Only 2 percent overall of respondents reported that someone at home went online at a community center, which afforded Internet access primarily to householders who were: 15 to 24 years old (4 percent), unemployed (4 percent), or American Indian or Alaska Native (4 percent). Ten percent of households reported that someone at home went online at an Internet café or coffee shop. The most frequent users there were: 15- to 24-year-olds (14 percent), employed (12 percent), high income earners at or above $100,000 annually (18 percent), college educated or better, (17 percent), and Asian American householders (17 percent).

Overall, 14 percent of households reported that someone in the home went online at someone else's house. Again, this number was highest among the youngest age brackets (15-24:22 percent; 25-44:20 percent) and those making $100,000 or above (21 percent). Those least likely to go online at someone else's house were those 65 or older (6 percent), individuals not in the labor force (9 percent), the disabled (8 percent), and individuals without a high school diploma (7 percent).

ConnectED

Broadband has tremendous potential to enhance education through customized lessons, rich digital content, and access to unprecedented amounts of information. However, less than 30 percent of schools have the high-speed Internet capacity necessary to access these benefits. In June 2013, President Obama announced ConnectEd, an effort to ensure that 99 percent of schools and libraries are connected to broadband with speeds of at least 100 Mbps with the goal of 1 Gbps by 2017. The ConnectEd initiative also seeks to fund teacher technology training and professional development to support the adoption of digital learning tools. As of February 2014, the private sector had committed $750 million in software, hardware, Internet connectivity, and educational tools and content to support the initiative. The Federal Communications Commission has directed $2 billion of funding from its E-Rate universal service program for schools and libraries be used as a down payment to extend broadband to 2o million students. In addition, the U.S. Department of Agriculture committed another $10 million for distance learning grants to rural schools. For a fact sheet describing ConnectED's progress, visit http://www.whitehouse.gov/the-press-office/2014/02/04/fact-sheet-opportunity- all-answering-president-s-call-enrich-american-ed.

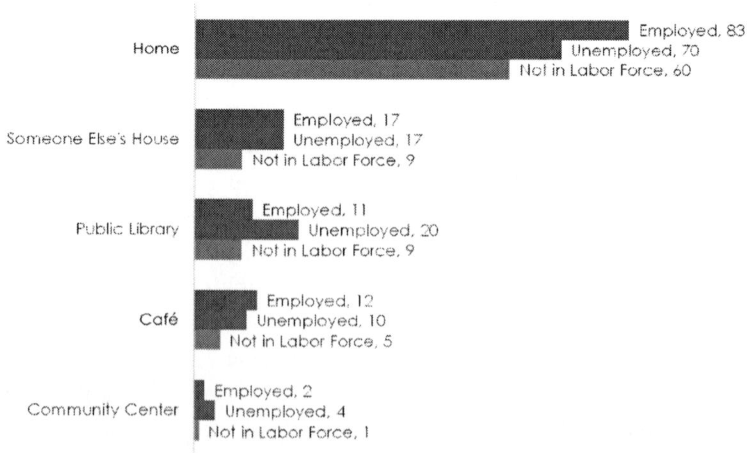

Figure 15. Internet Use Location by Householder Employment Status, Percent of Households, 2012.

Internet Use Anywhere (At-home or Away)

Internet use anywhere (either at home or away from home) continued to show positive upward movement in 2012. Since 2000, Internet use anywhere has grown by 31 percentage points to 75 percent of all Americans in 2012. The age brackets that reported the highest numbers in this category were the 15- to 24-year-olds (87 percent) and 25- to 44-year-olds (85 percent). The highest bump in Internet use anywhere occurred among those 65 and above, reporting a 37 percentage-point gain between 2000 and 2012. Adoption rates for using the Internet anywhere tended to rise along with income and education level, suggesting that persons with lower levels of family income and educational attainment may still lack resources and opportunities to get connected. Although in the year 2000 Hispanics were less likely than other ethnic groups to go online anywhere (24 percent), they closed the gap in 2012, by reporting numbers similar to other minority groups (Hispanic: 62 percent; American Indian and Alaska Native: 62 percent; African American: 68 percent). Persons living in rural areas still lagged behind their urban counterparts in Internet use anywhere, although both groups made over a 16 percentage-point gain since 2003 (rural: 69 percent; urban: 76 percent).

No Internet Use at Home

Even as Americans increasingly access the Internet through mobile devices and outside of the home, the Administration continues to encourage home broadband adoption. As the Internet evolves, some uses—such as consulting by video conference with a health professional or completing an online application with personal information—may require the privacy afforded by a home broadband connection. In addition, residential broadband connections can extend student learning beyond the classroom and facilitate digital literacy among household members unfamiliar or uncomfortable with Internet use. Families with broadband connections and telework options may have the ability to spend less on child care services by shortening commuting times. Finally, the advent of the "Internet of Things," which is now allowing consumers to monitor their energy consumption and home security systems remotely, among other activities, could accelerate the demand for broadband-enabled homes and the necessity of broadband adoption by all their occupants.[7]

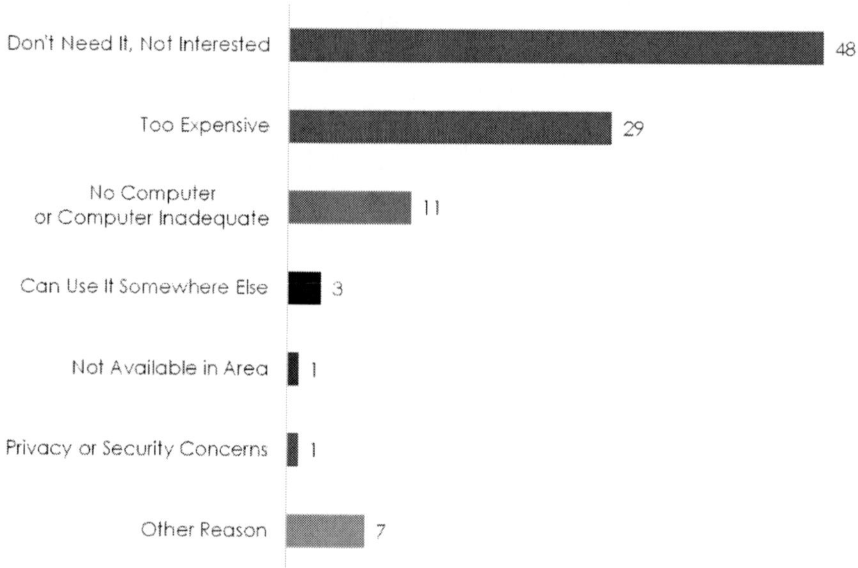

Figure 16. Main Reason for Non-Use of the Internet at Home, Percent of Households Not Online at Home, 2012.

As more Americans go online at home, a significant but declining number have yet to cross the digital divide by connecting to the Internet there. According to 2012 CPS data, 28 percent of the 122 million households represented in the survey did not use broadband at home. Including dial-up service decreases the proportion of households not online at home to 25 percent. Even so, households without Internet service at home offered a variety of primary reasons why they did not use it where they live (see Figure 16). Additionally, 18 percent of U.S. households did not use the Internet at all, regardless of location. The following discussion of the main reasons why some households declined to access the Internet at home, in order of their prevalence among 2012 CPS respondents, may assist policymakers as they pursue universal broadband adoption and affordable connectivity in every community in the nation.

No Need or Interest

In 2012, as in previous years, the prevailing reason that households without residential Internet service gave for not using it was the lack of need or interest in going online at home. In the 15 months between the July 2011 and October 2012 data collections, the portion of non-Internet using households offering this explanation remained at 48 percent. The percentage of such households increased 9 points between 2003 and 2012 from 39 percent to 48 percent.

Views on whether households needed Internet access or were interested in it varied based on previous home Internet use. Households that once used the Internet at home, but no longer did so as of the 2012 CPS, expressed disinterest in home Internet use much less frequently (21 percent) than the households that had never connected to the Internet from home (53 percent). Additionally, 38 percent of households that reported only using dial-up Internet service at home cited a lack of need for, or interest in, home broadband connections in 2012, an increase from 34 percent of dial-up users in 2011.

Dividing households by householder age shows differences in perceptions of whether home Internet use is needed or of interest. From 2001 to 2012, CPS data reveal a consistent pattern of non-adopting householders expressing no interest in, or no need for, home Internet service increasing with age. Accordingly, in 2012 non-Internet using householders ages 15 to 24 (24 percent), and those between 25 and 44 (26 percent), were the least likely groups to state they did not want or need to use the Internet at home. Forty-four percent of the next oldest age group, 45- to 64-year-olds, stated they did not want or need to use the Internet at home in 2012, and 70 percent of seniors ages 65 and older responded similarly. Between 2011 and 2012, however, disinterest increased among senior citizens over 65 years old (67 percent to 70 percent). While the incidence of senior non-using householders reporting they did not want or need to access the Internet at home rose 3 percentage points from 2011 to 2012, this increase followed a significant 11 percentage-point jump from 57 percent to 68 percent between the 2009 and 2010 CPS data collections.

In recent years, employers have more actively recruited through online job postings that extend beyond the declining circulation of some local newspapers or the limited reach of community job banks (Salpeter, 2012). CPS data from 2001 through 2012 demonstrate that unemployed householders were always the least likely to state they did not need to use the Internet at home. For example, in 2012, non-Internet using households led by an unemployed person expressed disinterest in home Internet use (19 percent) less often than households headed by employed individuals (39 percent) or those "not in the labor force" (58 percent).[8] Between 2011 and 2012, unemployed-led households responding it was not necessary or desirable to go online at home declined 3 percentage points from 22 percent, while the percentage of households headed by working householders giving the same reason remained unchanged. At the same time, households of those outside the labor force continued to exhibit the least interest in home Internet adoption, showing a 1 percentage-point increase from 57 percent in 2011 to 58 percent in 2012.

In 2012, the most affluent non-using households, with annual incomes of $100,000 or more, stated less often than other income groups that they did not want or need to use the Internet in their homes (45 percent), followed closely by households with family incomes below $25,000 per year (47 percent). About half of households with annual incomes ranging between $25,000 and $99,999 expressed disinterest in accessing the Internet at home (see Figure 19). Comparing data from the 2011 CPS, the lowest income households, those earning below $25,000 annually, were the least likely to offer a lack of need or interest in using the Internet at home (45 percent), while households with incomes at or above $100,000 were the most likely to say they did not need to do so (53 percent). Between 2011 and 2012, disinterest in residential Internet use dropped 8 percentage points among these highest income households.

	2001	2003	2009	2010	2011	2012
No Diploma	55	40	40	48	49	50
High School Diploma	53	41	41	50	50	50
Some College	46	36	35	41	41	43
College Degree	47	41	40	45	48	50

Note: Educational attainment is an attribute of the householder.

Figure 17. Households Citing Lack of Interest as Main Reason for Not Going Online by Education, Percent of Households Not Online at Home, 2001-2012

Non-using householders with some college education or an associate degree were less likely to report not going online at home due to a lack of interest than others with higher or lower educational attainment levels. Forty-three percent said they did not get online at home because they were not interested in accessing the Internet there, up from 41 percent in 2011. In 2012, 50 percent of every other category of academic completion expressed no interest in connecting to the Internet where they lived. Except in 2001, when 8 percentage points separated the least educated (55 percent) and the most educated (47 percent) householders who said they did not need to visit the

Internet at home, the difference between these groups has been nominal (see Figure 17).

Following a pattern from 2001, more men than women expressed disinterest in going online at home in 2012, with 50 percent of male heads of non-using households and 47 percent of their female counterparts reporting this answer. The percentage of men offering this response increased 2 points from 2011, but was unchanged for women at 47 percent.

	2001	2003	2009	2010	2011	2012
White	56	46	45	53	53	55
African American	40	29	33	40	40	40
Hispanic	39	25	30	35	39	38
Asian American	43	36	39	46	56	49
American Indian or Alaska Native	38	31	34	37	34	31
Other	N/A	30	28	35	39	38

Note: Race and ethnicity are attributes of the householder.

Figure 18. Households Citing Lack of Interest as Main Reason for Not Going Online by Race, Percent of Households Not Online at Home, 2001-2012.

The race or ethnicity of a householder correlates with whether a household not online is interested in using the Internet while home. Between 2001 and 2012, White households have most often stated no interest or need to connect to the Internet in their homes, except in 2011, when 56 percent of Asian American non-using householders stated no desire to do so, compared to 53 percent of Whites. In all other years, Asian American householders were the next group after Whites to express disinterest in using the Internet at home. In the period between 2009 and 2010, the proportion of households of all racial and ethnic groups reporting they lacked interest in, or need for, home Internet use increased markedly, although less so among American Indians and Alaska Natives. From 2011 to 2012, these households were the least likely to say they were not interested in accessing the Internet from home, with the percentage of such responses declining 3 points during the period.

In 2010, Hispanic households (35 percent) were the least likely to say they did not need to use the Internet where they lived, but disinterest expressed by this group increased to 39 percent in 2011, before falling slightly to 38 percent in 2012. The percentage of American Indian and Alaska Native households that stated they were not interested in home Internet access decreased from 37 percent in 2010 to 34 percent in 2011 to 31 percent in 2012. While the percentage of African American households expressing that view remained flat at 40 percent from 2010 to 2012, Asian American households offering disinterest to explain why they did not use the Internet at home increased 10 percentage points between 2010 and 2011 (2010: 46 percent; 2011: 56 percent) then dropped 7 points to 49 percent between 2011 and 2012. (see Figure 18).

Generally, between 2001 and 2012, rural residents stated they did not need to use the Internet at home more often than urban residents. CPS data for 2012 show 52 percent of rural residents explained they were not interested in home Internet use, while 48 percent urban residents replied they were uninterested in home Internet).

As in past years, households with school-age children were more likely to consider home Internet use an important supplement to classroom education than households without children in school. In 2001, 49 percent of non-using families with children stated they lacked a need or interest in accessing the Internet at home compared to 57 percent of households without children, but 2012 CPS data reflect decreased disinterest among both groups at 47 percent and 54 percent respectively. Interestingly, between 2009 and 2010, expressions of disinterest in home Internet use increased significantly among families with school-age children (38 percent to 46 percent) and those without them (45 percent to 52 percent).

A higher incidence of U.S. householders who were disabled cited their lack of need or interest in home Internet connections at 56 percent, compared to the 48 percent of all U.S. households that did not access the Internet at home in 2012. Yet in 2011, 51 percent of people with disabilities explained they did not visit the Internet at home because they did not need to. This figure represented a 3 percentage-point decline from 2010 (54 percent) and suggests that further research could help shed light on the reluctance of some people with disabilities to use residential Internet services despite the significant potential benefits such access could provide. For example, in 2012, 61 percent of deaf or hard of hearing individuals who did not access the Internet at home expressed the highest disinterest among people with various disabilities. These respondents may, however, be unaware of, or unable to afford, assistive technologies that could facilitate their use of the Internet at home.

Expense

The expense of using the Internet at home remained the second most often cited reason non-Internet households offered in 2012 as the main reason why they did not connect there. Although households indicating a lack of interest exceed those naming expense as the main obstacle, the number of households citing expense should raise concerns for policymakers. For those households—7 percent of all American households—high costs or low income may present significant barriers to going online.

	No Need or Interest	Too Expensive	No/ Inadequate Computer
All Households Not Online	48	29	11
Family Income			
Less Than $25,000	47	32	11
$25,000-49,999	51	26	11
$50,000-74,999	52	22	10
$75,000-99,999	51	20	5
$100,000 or More	45	19	10
Education*			
No Diploma	50	30	12
High SchoolDiploma	50	28	11
Some College	43	31	11
College Degree	50	24	9
Race and Ethnicity*			
White	55	22	10
African American	40	37	13
Hispanic	38	41	12
Asian American	49	26	11
American Indian or Alaska Native	31	42	8
Population Density			
Urban	48	30	11
Rural	52	25	10

* These are attributes of the householder.

Figure 19. Main Reason for Non-Use of the Internet at Home by Demographics and Population Density, Percent of Households Not Online at Home, 2012.

Twenty-nine percent of unconnected households responded in 2012 that they viewed the cost of going online at home as too high. The proportion of households citing expense has since grown steadily, from 23 percent in 2003 to 28 percent in 2011. Among households that ceased using the Internet at home by 2012, the expense of such service was the most frequent explanation for why they had relinquished it (43 percent), and they gave this response twice as often as their lost need or interest in home Internet use (21 percent). In contrast to formerly-using households, those that had never been online at home stated disinterest twice as often (53 percent) as too expensive (27 percent). Only 2 percentage points separated dial-up households expressing disinterest in high-speed home Internet connections (38 percent) from such households responding that cost prevented them from using residential broadband service (36 percent) in 2012.

Not surprisingly, non-using households led by the youngest householders ages 15 to 24 most often stated that high cost prevented their Internet use at home, rising since 2001 to a high of 51 percent in 2011, but declining a percentage point in 2012 to 50 percent. In 2012, more than twice as many of these younger householders explained that cost was the main barrier to home Internet use compared to the 24 percent of such householders who expressed disinterest in home Internet adoption. Forty-nine percent of householders ages 25 to 44 also reported expense was the primary deterrent to their home Internet use. Householders 65 years of age or older were consistently the least likely age group to cite expense as the primary reason they did not go online at home, and the most likely to respond they did not desire to connect to the Internet there. In 2001, 11 percent of seniors said they thought residential Internet service was too expensive, while 72 percent said they did not want it. By 2012, the percentage of this group concerned about expense remained at 11 percent and those citing no interest or need declined 2 percentage points to 70 percent from 72 percent in 2001.

Predictably, householder employment status affected the likelihood that respondents raised concerns about the expense of home Internet use, and unemployed householders were the most likely to do so. With employers rapidly migrating to online job postings, the jobless increasingly require Internet access to search and train for work but can least afford to go online from home. In 2012, 58 percent of unemployed persons reported that expense was the main reason for not using the Internet at home, which represented the highest percentage of respondents offering this reason between 2001 and 2012. Further, in the 15 months between the July 2011 and October 2012 data collections, a 3 percentage-point increase occurred among both unemployed

(55 percent to 58 percent) and employed householders (33 percent to 36 percent) who replied the high cost of using the Internet where they live prevented them from obtaining access. Among those not in the workforce who responded similarly, their proportion rose 4 percentage points from 17 percent in 2009 and 2010 to 21 percent in 2011 and 2012.

CPS data reveal an inverse relationship between family income and the incidence of responding households to reply that high Internet costs prevented them from getting online in their homes. Consequently, the lowest earning households most often stated that financial concerns were the primary factor for not having residential Internet service. Both in 2012 and 2011, 32 percent of households with family incomes less than $25,000 cited expense, while 19 percent of families earning $100,000 or more annually gave the same reason in each of those years. Families with household incomes between $25,000 and $49,999 expressing concern about the expense of accessing the Internet at home remained constant at 26 percent from 2011 to 2012, but grew 4 percentage points among those earning between $50,000 and $74,999 during that period. In 2012, non- adopting households earning less than $25,000 annually were the income group most concerned about the cost of home Internet service, but significantly, they were also the least likely to say they had no interest or need for such service (see Figure 19).

Each year from 2009 to 2012 (27 percent in 2009, 28 percent in 2010, 32 percent in 2011, and 31 percent in 2012), householders with some college or an associate degree were the most likely to say the Internet's expense was the main reason they did not use the service at home. By comparison, during that same period householders without a high school diploma expressed that concern less frequently (25 percent in 2009 and 2010, 28 percent in 2011, and 30 percent in 2012). Since 2001, householders who have earned at least a college degree consistently have been the least likely to cite expense as their primary reason for declining to use the Internet at home. Significantly, the proportion of highly educated householders deterred from going online at home because of the high cost has climbed in recent years from 18 percent in 2009 to 20 percent in 2010, to 23 percent in 2011, reaching a peak of 24 percent in 2012.

By 3 percentage points, female householders (30 percent) were more likely than their male counterparts (27 percent) to explain that expense prevented them from using the Internet at home in 2012. Such responses increased among both genders by a single percentage point from 2011. The frequency with which both groups cited the high cost of accessing the Internet at home increased among male and female householders between 2010 and 2011, from 22 to 26 percent for men and from 26 to 29 percent for women.

CPS data continue to show that affordability as a household's main reason for not using the Internet at home varied among racial and ethnic groups. Since 2001, Whites and Asian Americans have been the least likely to cite expense as their main impediment to home Internet use. Further, White households' responses have varied the least between 2001 and 2012, and were unchanged from the 2011 CPS at 22 percent. Between 2011 and 2012, the percentage of African American householders offering this reason dropped 1 point to 37 percent. At the same time, the percentage of all other groups reporting expense as the most important reason for not using home Internet connections increased between 2011 and 2012. Hispanic households providing the same response rose a modest 4 percentage points from 37 percent to 41 percent compared to the 9 percentage-point gain reported by Asian American householders (17 percent to 26 percent) and the dramatic 17 percentage-point jump reflected in the responses of American Indians and Alaska Natives (25 percent to 42 percent) from 2011 to 2012. During the preceding period between 2010 and 2011, the incidence of African American householders citing financial concerns increased by 8 percentage points from 30 percent to 38 percent. Yet the proportion of Hispanic householders deterred by cost considerations from going online at home grew more slowly by 2 percentage points from 35 percent to 37 percent between 2010 and 2011.

In 2003, African American (33 percent to 29 percent) and Hispanic householders (33 percent to 25 percent) cited expense over lack of need as the main reason for not using the Internet at home. Hispanics continued in 2009 to identify cost concerns as the most important reason for not going online at home, as opposed to not wanting or needing to do so (34 percent to 30 percent). A different pattern began to emerge among Hispanic householders in 2010, when both reasons tied at 35 percent, and by 2011 their responses began to diverge, with disinterest at 39 percent surpassing expense at 37 percent. By 2012, however, the expense of home Internet use (41 percent) rather than disinterest (38 percent) again predominated Hispanic householders' reasons for not going online there. Finally, the sharp increase during this period from 25 percent to 42 percent of American Indian and Alaska Native households citing expense as their main concern also resulted in that reason outranking disinterest (31 percent) among this group for the first time (see Figures 18 and 20).

Consistent with historical patterns, in 2012 rural households (25 percent) expressed less concern about the expense of home Internet use than did urban households (30 percent). Between 2011 and 2012, urban respondents who stated online access at home was too expensive rose by a percentage point, but the frequency of that response did not change among rural households. The percentages citing expense for both groups grew from 2010 (21 percent rural; 25 percent urban) to 2011 (25 percent rural; 30 percent urban).

Once again, household responses about why they did not use the Internet at home varied based on the presence of school-age children. In 2012, as in past years, families with school children (30 percent) were more likely than those without (24 percent) to say expense primarily explained why they did not use home Internet connections. In addition, 22 percent of householders with a disability expressed cost concerns as their primary reason for not going online at home.

No or Inadequate Computer

Since 2003, the frequency of this reason for no Internet use at home has declined over time from 23 percent that year to a low of 11 percent in 2012, falling 2 percentage points from 2011. According to the 2012 CPS, only 1 percent of dial-up households cited no computer or one that did not work sufficiently well for their purposes to explain why they chose not to use more robust broadband connections at home. Yet 18 percent of households that no longer used the Internet at home and 10 percent of homes that had never been online explained they lacked a computer or an adequate one.

	2001	2003	2009	2010	2011	2012
White	19	18	19	20	22	22
African American	38	33	30	30	38	37
Hispanic	39	33	34	35	37	41
Asian American	27	22	20	18	17	26
American Indian or Alaska Native	34	28	25	22	25	42
Other	N/A	27	33	31	41	41

Note: Race and ethnicity are attributes of the householder.

Figure 20. Households Citing Expense as Main Reason for Not Going Online by Race, Percent of Households Not Online at Home, 2001-2012.

Except in 2003, when 10 percentage points separated the youngest respondents ages 15 to 24 (29 percent) from seniors 65 years or older (19 percent), age did not seem to affect this response as the main reason householders of various age groups did not use the Internet at home. In 2011, 14 percent of householders ages 25 to 44 replied they had an unsatisfactory computer or none at all, overtaking by 3 percentage points the 11 percent of 15- to 24-year-olds who offered that explanation. The portion of all age groups citing this reason declined over time and by 2012, these two youngest age groups tied at 12 percent, followed by 45- to 64-year-olds at 11 percent, and seniors ages 65 or older at 10 percent.

Householder employment status apparently has no effect on this reason for no home Internet use. In 2011, 13 percent of both unemployed householders and those not in the labor force cited no computer or a poor one for not going online at home, while 12 percent of employed respondents gave the same reason. In 2012, 11 percent of employed householders, as well as householders outside the labor force gave this response, as did 12 percent of unemployed CPS participants.

Similarly, breaking out households by family income did not reveal large differences in the frequency with which respondents stated they did not go online at home because they lacked an adequate computer.[9] In 2011, only 3 percentage points separated the highest income families earning at least $100,000 annually (10 percent) from the lowest income household earning less than $25,000 per year (13 percent) responding that no computer or an inadequate one kept them offline at home. Only a single percentage point differentiated households making as much as $49,999 (11 percent) from households with annual incomes from $50,000 to $74,999, as well as those earning $100,000 or above (10 percent). Interestingly, in 2012, families earning $75,000 to $99,999 were half as likely as other income groups to respond that no or poor computer equipment prevented them from accessing the Internet at home (5 percent) (see Figure 19). CPS results for 2012 showed a large 8 percentage-point drop from 13 percent in 2011 among households in this $75,000 to $99,999 per year income range.

A householder's lower education level indicated a slightly higher incidence of no computer or an inadequate one as the main reason the household did not use residential Internet service. Between 2003 and 2012, a 4 percentage-point gap existed between householders with a college degree or more and those without a high school diploma generally, except in 2009, when the difference reached 7 percentage points (16 percent to 23 percent). The difference declined to 3 percentage points in 2012 (9 percent to 12 percent). In

2012, the frequency of this response dropped 2 percentage points from 2011 among those without a high school diploma (14 percent to 12 percent) and high school graduates (13 percent to 11 percent). From 2011 to 2012, the percentage gap fell to only 1 point among those with some college or an associate degree (12 percent to 11 percent) and those with at least college degree (10 percent to 9 percent).

American Indian and Alaska Native respondents were the least likely, at 8 percent, to say they lacked any or an adequate computer as the most important reason for not using the Internet at home. In 2010, 23 percent of American Indians and Alaska Natives compared to 14 percent of Whites gave that reason, with 17 percent each of African American and Hispanic householders and 18 percent of Asian American householders citing the same response. In 2012, 10 percent of White householders said their poor or absent computer kept them offline at home, as Asian American (11 percent), Hispanic (12 percent), and African American (13 percent) householders offered similar explanations for their non-use (see Figure 19). Data from all groups show a decrease in the incidence of this response between 2011 and 2012 (including among American Indians and Alaska Natives—18 to 8 percent; Hispanics—15 to 12 percent; Asian Americans—14 to 11 percent; and Whites—13 to 10 percent), except among African American householders, which grew from 11 percent to 13 percent during the period.

Householder gender made little difference between male and female respondents stating they lacked any or an adequate computer to explain why they did not access the Internet from home. In 2011, 13 percent of each gender stated this reason, which dropped to 11 percent for both men and women in 2012. Two percentage points separated men and women in 2009, with men less likely (20 percent) than women (22 percent) to cite access to an insufficient computer, if one was available at all, to explain why they did not use the Internet at home.

Rural or urban residence also negligibly affected whether a household identified no computer or an inadequate one as the major reason for no residential Internet use. Ten percent of rural householders compared to 11 percent of urban householders offered this reason in 2012, with each decreasing from 13 percent in 2011.

As the frequency of households citing computer inadequacy declined over time, only one percentage point distinguished households with school children and families without them in 2003 (23 percent to 22 percent), 2009 (21 percent to 20 percent), 2010 (15 percent to 16 percent), and 2012 (11 percent to 10 percent). In 2011, these two household types tied at 13 percent in the respective frequency of this explanation for no home Internet use. Among people with disabilities, 10 percent expressed dissatisfaction with their computer's adequacy or stated they had no computer to use for Internet access at home in 2012.

Other Reasons

The proportion of households offering this response for why they did not have the Internet at home remained unchanged at 7 percent since 2010, up slightly from 6 percent in 2009. As 2012 data show, 6 percent of households that have never had home Internet access cited "other" reasons for not connecting at home, compared to dial-up households at 9 percent and formerly-Internet-using households at 10 percent.

Can Use It Elsewhere

In 2012, households stating they did not access the Internet at home because they could go online at some other location remained constant at 3 percent from 2011, down from 5 percent in 2009 and 2010. Only 1 percent of dial-up households stated they could use the Internet at a different location in 2012, compared to 3 percent of households that have never had residential Internet service and 4 percent of households that previously used the Internet at home.

Consistently, householders ages 15 to 24 and 25 to 44 were the most likely to say they could access the Internet at a location other than their own homes, possibly because service was available to them at school or work. Seniors 65 years or older were predictably the least likely to state alternative access as their main reason. For example, in 2012, 5 percent of 15- to 44-year-olds gave this as their main reason, followed by householders ages 45 to 64 at 4 percent, and those 65 and older at 1 percent.

More employed and unemployed individuals offered this reason, compared to householders not in the labor force. Conceivably, employed householders could use the Internet where they work, just as unemployed household heads may find access at job placement and training sites or other locations. Data from the 2012 CPS demonstrate that employed householders (5 percent) were more than twice as likely as those not in labor force (2 percent) to give this response. This represents a decrease from 2011 among employed

(6 percent) and unemployed (3 percent) non-using householders, but doubled to 2 percent in 2012 for those not in labor force.

Income appears to be positively related to the opportunity to use the Internet outside of the home, with the lowest earning households least likely to give alternative location as the primary reason they did not use the Internet at home. In 2012, householders earning $75,000 to $99,999 (9 percent) were three times more likely to give this answer than those earning $25,000 to $49,999 (3 percent). The most affluent householders with family incomes of at least $100,000 (8 percent) were four times more likely than those earning less than $25,000 (2 percent) to reply they could use the Internet somewhere else. The same pattern held in 2011, and remained constant for the highest and lowest income households, but increased from 5 percent to 9 percent in 2012 for households earning $75,000 to $99,999 per year.

As with income, the likelihood that respondents replied they had other places to access the Internet increases with educational attainment. The proportions did not fluctuate between 2011 and 2012 for householders with: no diploma (1 percent), a high school diploma (3 percent), some college courses or associate degree (4 percent), and a college degree or higher (7 percent). The largest percentage-point change reported by higher educated householders occurred between 2010 and 2011, when the incidence dropped 4 percentage points from 8 percent to 4 percent among those with some college, and by 5 points from 12 percent to 7 percent among householders with at least a college degree.

Male and female households responding they could use the Internet elsewhere held firm for each at 3 percent in 2011 and 2012. Two percent of individuals with disabilities reported they could go online at another location in the 2012 CPS.

CPS data indicate small differences in householders explaining they could go elsewhere to connect to the Internet based on their race and ethnicity. In 2003, "Others" provided this response most often at 4 percent, followed by Whites and Asian Americans at 3 percent each, American Indians and Alaska Natives at 2 percent, and African Americans and Hispanics at 1 percent each. However, the proportion of all racial and ethnic groups offering this as the main reason for no home Internet has increased over time between 2003 and 2010, peaking in 2010 with American Indian and Alaska Native householders at 9 percent, African American householders at 6 percent and White, Asian American, and Hispanic householders each at 5 percent. The percentages of American Indian and Alaska Native householders fell by two-thirds in 2011 to 3 percent, before doubling again to 6 percent in 2012, when again they were the most likely to say their ability to use the Internet at an alternative location explained why they did not connect to the Internet where they lived. Between 2011 and 2012, the portion of Whites giving this reason remained stable at 3 percent as did that of Asian Americans at 4 percent, but African American and Hispanic householder responses dropped to 3 percent and 2 percent, respectively.

In 2011, rural (2 percent) and urban (3 percent) respondents reported other locations as their main reason for no home Internet use, but they tied each at 3 percent in 2012. No difference existed between the responses of non-using families with or without school- age children, with 5 percent of each responding in this fashion in 2010, compared to 3 percent each in 2011 and 2012.

Not Available in Area

This response has remained constant for both rural and urban householders from 2009 to 2012. During that time, rural residents were twice as likely at 2 percent to explain they did not connect to the Internet at home because no Internet service was available where they lived compared to 1 percent of urban residents. American Indian and Alaska Native households were the most likely racial group to give this reason, with the ratio fluctuating from 3 percent in 2009, to 4 percent in 2010, up significantly to 11 percent in 2011, before settling back to 4 percent in 2012. Three percent of non-using households with family incomes of $100,000 or more reported that residential Internet service was unavailable, compared to 1 percent of those households with incomes below $50,000. Fifteen percent of dial-up households reported they did not use high-speed connections at home because residential broadband service was inaccessible to them.

Privacy

Although only 1 percent of households expressed privacy concerns in both 2011 and 2012 as their primary reason for not using the Internet at home, well-publicized data breaches and greater consumer awareness of Internet privacy issues may affect this response in future years. In view of its ongoing efforts to advance policies to protect consumer data privacy, NTIA will carefully monitor upcoming data on this question.

As households increase their home Internet use, research exploring the value proposition for non-Internet consumers could help policymakers understand, for example, whether persistent non-users do not find such access integral to their daily lives and therefore are not interested in it, whether they do not find sufficient utility in home Internet use to pay the costs they perceive as too high, or whether they truly desire residential Internet service, but simply cannot afford it. Further research on these and other questions could help government and community leaders best tailor policies that facilitate sustained broadband adoption by Americans who cannot use this empowering technology or have not yet to chosen to do so.

CONCLUSION

From the start of the Digital Nation series and predecessor reports on the digital divide beginning in 1994, each survey provides new insight into how Americans are accessing and using the Internet. NTIA looks forward to further exploring the questions generated by the current data, while continuing to find new ways to analyze and report findings to best meet the needs of social scientists and policymakers. In the next Digital Nation report, NTIA hopes to develop a greater understanding of why Americans choose to go online, and the device choices they make to do so. As NTIA uses this information to advance national policy to reach the critical goal of making the Internet available and affordable for every American, we urge each community to accelerate local efforts to bridge the digital divide through actively engaging citizens and elected officials, as well as its civic, business, and non-profit leaders. Our nation's future depends in part on our success in this common pursuit.

APPENDIX A. DATA AND METHODOLOGY

This report uses data from the October 2012 Current Population Survey ("CPS"), a monthly survey of a representative sample of the U.S. noninstitutional population that provides data on labor force participation, income, and demographic characteristics of households. It includes data from the October 2012 CPS Computer and Internet Use Supplement, a special supplement to the CPS periodically commissioned by NTIA.

The October 2012 CPS asked each surveyed household whether someone in that household used or owned a computer, as well as who in the household used the Internet, and the devices and locations from which they did so. In addition, the survey asked the household which of the following technologies members utilized to connect to the Internet from home: dial-up service, DSL, cable modem, fiber optics, satellite, mobile broadband, or some other Internet connection technology. Using these data, it is possible to determine whether a household owned a personal computer (i.e., desktop, laptop, netbook, or notebook) or a handheld device, as well as the type of broadband technology (i.e., DSL, cable modem, fiber optics, satellite, or mobile broadband) members utilized to connect to the Internet. The survey also asked those households that did not access Internet services to state their main reason for not doing so.

The October 2012 CPS Supplement included a group of questions primarily concerning Internet and mobile phone usage habits.[10] These questions were asked of the household respondent about his/her own usage habits. Since one person in each household responded and that person was not randomly selected, a special set of weights was created to properly tally these variables. Care should be used when analyzing these data because the respondents are not evenly distributed across the sample based on age. This is why analysis of these data only includes the population ages 25 and older.

About 53,600 household records comprise the sample, representing 122 million American households. NTIA analyzes computer and Internet use at the household and person levels and their association with characteristics such as age, family income, household size and composition, and geographic location. When conducting household- level analysis, the authors use information for the household reference person (also known as the "head of household" or "householder") as proxies for characteristics like education, race, ethnicity, age, disability status, and foreign-born status.

Prior to October 2010, data on computer use, as well as the types of broadband technology that online households utilized, had not been available since the early 2000s. The supply and demand for both mobile devices and residential Internet access services have changed enormously during this period. The October 2012 data serve to update these trends. The data from the October 2012 CPS make it possible to continue to identify the preferred or most common types of computers and access technologies used for residential Internet access, including the prevalence of mobile broadband technologies and handheld devices.

APPENDIX B. STATISTICAL TABLES

Table 1. Regression of Mobile Phone-Based Email Use on Demographic and Geographic Characteristics, Mobile Phone Users Age 25+, 2012

Variable	Coefficient Estimate	Standard Error
Family Income: $25,000-49,999	0.0412 **	0.0077
Family Income: $50,000-74,999	0.0851 **	0.0071
Family Income: $75,000-99,999	0.1333 **	0.0096
Family Income: $100,000 or More	0.2239 **	0.0090
Education: High School Diploma	0.0581 **	0.0096
Education: Some College	0.1561 **	0.0099
Education: College Degree or More	0.2207 **	0.0103
Age	-0.0196 **	0.0011
Age Squared	0.0001 **	0.0000
Race: African American	-0.0034	0.0088
Race: Hispanic	-0.0029	0.0096
Race: Asian American	-0.0135	0.0130
Race: American Indian or Alaska Native	0.0335	0.0345
Race: Other	0.0183	0.0192
Female	-0.0047	0.0050
Disabled	-0.0122	0.0083
Foreign-Born Non-Citizen	-0.0696 **	0.0130
Employment Status: Unemployed	-0.0608 **	0.0124
Employment Status: Not in Labor Force	-0.0577 **	0.0062
Related School-Age Children at Home	0.0725 **	0.0177
Household Size	-0.0057	0.0035
Household Size When Related School-Age Children at Home	-0.0100 †	0.0051
Metropolitan Status: Urban	0.0847 **	0.0081
Metropolitan Status: Unidentified	-0.0267	0.0449
Region: Midwest	0.0112	0.0083
Region: South	0.0514 **	0.0080
Region: West	0.0719 **	0.0079
Constant	0.8397 **	0.0310
Age 25+ Sample Size	50,627	
Estimated Age 25+ Population Size	205,182,829	
Mobile Phone User Subsample Size	44,019	
Estimated Mobile Phone User Subpopulation Size	180,320,465	
R^2	0.2134	

† indicates 90 percent confidence that the marginal effect is nonzero ($p \leq 0.1$).
* indicates 95 percent confidence that the marginal effect is nonzero ($p \leq 0.05$).
** indicates 99 percent confidence that the marginal effect is nonzero ($p \leq 0.01$).

Table 2. Regression of Mobile Phone-Based Web Browsing on Demographic and Geographic Characteristics, Mobile Phone Users Age 25+, 2012

Variable	Coefficient Estimate	Standard Error
Family Income: $25,000-49,999	0.0290 **	0.0069
Family Income: $50,000-74,999	0.0688 **	0.0078
Family Income: $75,000-99,999	0.1068 **	0.0102
Family Income: $100,000 or More	0.1885 **	0.0093
Education: High School Diploma	0.0522 **	0.0088
Education: Some College	0.1365 **	0.0101
Education: College Degree or More	0.1846 **	0.0111
Age	-0.0218 **	0.0011
Age Squared	0.0001 **	0.0000
Race: African American	0.0038	0.0081
Race: Hispanic	-0.0217 *	0.0096
Race: Asian American	-0.0350 **	0.0132
Race: American Indian or Alaska Native	0.0296	0.0353
Race: Other	0.0168	0.0222
Female	0.0064	0.0048
Disabled	-0.0087	0.0080
Foreign-Born Non-Citizen	-0.0641 **	0.0126
Employment Status: Unemployed	-0.0441 **	0.0126
Employment Status: Not in Labor Force	-0.0505 **	0.0065
Related School-Age Children at Home	0.0485 **	0.0186
Household Size	-0.0019	0.0032
Household Size When Related School-Age Children at Home	-0.0054 †	0.0051
Metropolitan Status: Urban	0.0579 **	0.0082
Metropolitan Status: Unidentified	-0.0489	0.0516
Region: Midwest	0.0343	0.0084
Region: South	0.0513 **	0.0081
Region: West	0.0803 **	0.0084
Constant	0.9484 **	0.0310
Age 25+ Sample Size	50,627	
Estimated Age 25+ Population Size	205,182,829	
Mobile Phone User Subsample Size	44,019	
Estimated Mobile Phone User Subpopulation Size	180,320,465	
R^2	0.2061	

† indicates 90 percent confidence that the marginal effect is nonzero ($p \leq 0.1$).

* indicates 95 percent confidence that the marginal effect is nonzero ($p \leq 0.05$).

** indicates 99 percent confidence that the marginal effect is nonzero ($p \leq 0.01$).

Table 3. Regression of Mobile Phone-Based App Downloading on Demographic and Geographic Characteristics, Mobile Phone Users Age 25+, 2012

Variable	Coefficient Estimate	Standard Error
Family Income: $25,000-49,999	0.0396 **	0.0067
Family Income: $50,000-74,999	0.0747 **	0.0071
Family Income: $75,000-99,999	0.1187 **	0.0095
Family Income: $100,000 or More	0.1845 **	0.0087
Education: High School Diploma	0.0390 **	0.0091
Education: Some College	0.1115 **	0.0082
Education: College Degree or More	0.1455 **	0.0099
Age	-0.0230 **	0.0010
Age Squared	0.0001 **	0.0000
Race: African American	-0.0090	0.0080
Race: Hispanic	-0.0371 **	0.0090
Race: Asian American	-0.0261 *	0.0128
Race: American Indian or Alaska Native	0.0035	0.0328
Race: Other	0.0199	0.0197
Female	-0.0002	0.0045
Disabled	-0.0062	0.0073
Foreign-Born Non-Citizen	-0.0815 **	0.0120
Employment Status: Unemployed	-0.0421 **	0.0112
Employment Status: Not in Labor Force	-0.0382 **	0.0055
Related School-Age Children at Home	0.0640 **	0.0177
Household Size	0.0021	0.0035
Household Size When Related School-Age Children at Home	-0.0138 **	0.0052
Metropolitan Status: Urban	0.0584 **	0.0078
Metropolitan Status: Unidentified	-0.0316	0.0335
Region: Midwest	0.0251 **	0.0084
Region: South	0.0438 **	0.0081
Region: West	0.0797 **	0.0077
Constant	0.8562 **	0.0312
Age 25+ Sample Size	50,627	
Estimated Age 25+ Population Size	205,182,829	
Mobile Phone User Subsample Size	44,019	
Estimated Mobile Phone User Subpopulation Size	180,320,465	
R^2	0.1725	

† indicates 90 percent confidence that the marginal effect is nonzero ($p \leq 0.1$).

* indicates 95 percent confidence that the marginal effect is nonzero ($p \leq 0.05$).

** indicates 99 percent confidence that the marginal effect is nonzero ($p \leq 0.01$).

Table 4. Regression of Mobile Phone-Based Social Networking on Demographic and Geographic Characteristics, Mobile Phone Users Age 25+, 2012

Variable	Coefficient Estimate	Standard Error
Family Income: $25,000-49,999	0.0352 **	0.0067
Family Income: $50,000-74,999	0.0651 **	0.0075
Family Income: $75,000-99,999	0.0779 **	0.0100
Family Income: $100,000 or More	0.1227 **	0.0088
Education: High School Diploma	0.0359 **	0.0084
Education: Some College	0.0928 **	0.0085
Education: College Degree or More	0.1168 **	0.0105
Age	-0.0274 **	0.0010
Age Squared	0.0002 **	0.0000
Race: African American	-0.0094	0.0086
Race: Hispanic	-0.0190 *	0.0087
Race: Asian American	-0.0226 †	0.0131
Race: American Indian or Alaska Native	0.0464	0.0324
Race: Other	0.0421 *	0.0203
Female	0.0482 **	0.0049
Disabled	0.0001	0.0067
Foreign-Born Non-Citizen	-0.0703 **	0.0115
Employment Status: Unemployed	-0.0468 **	0.0121
Employment Status: Not in Labor Force	-0.0342 **	0.0055
Related School-Age Children at Home	0.0640 **	0.0164
Household Size	0.0012	0.0031
Household Size When Related School-Age Children at Home	-0.0132 **	0.0048
Metropolitan Status: Urban	0.0341 **	0.0073
Metropolitan Status: Unidentified	-0.0869 *	0.0371
Region: Midwest	0.0311 **	0.0075
Region: South	0.0499 **	0.0071
Region: West	0.0608 **	0.0074
Constant	0.9967 **	0.0301
Age 25+ Sample Size	50,627	
Estimated Age 25+ Population Size	205,182,829	
Mobile Phone User Subsample Size	44,019	
Estimated Mobile Phone User Subpopulation Size	180,320,465	
R^2	0.1739	

† indicates 90 percent confidence that the marginal effect is nonzero ($p \leq 0.1$).

* indicates 95 percent confidence that the marginal effect is nonzero ($p \leq 0.05$).

** indicates 99 percent confidence that the marginal effect is nonzero ($p \leq 0.01$).

Table 5. Internet Use at Home by State, Percent of Households, 2009-2012

State	2012 95% Conf. Interval			2006 95% Conf. Interval			Est. % Chg. 09-12
	Est.	Lower	Upper	Est.	Lower	Upper	
Alabama	64.7	60.8	68.4	56.1	52.1	60.1	15.3
Alaska	78.1	74.7	81.2	77.3	73.9	80.3	1.0
Arizona	71.6	67.8	75.1	72.1	68.4	75.5	-0.7
Arkansas	68.8	63.5	73.7	55.9	51.8	59.9	23.1
California	78.2	76.6	79.7	73.1	71.7	74.4	7.0
Colorado	82.5	79.5	85.1	72.9	70.3	75.4	13.2
Connecticut	78.9	76.2	81.3	74.8	72.2	77.1	5.5
Delaware	74.1	70.7	77.3	71.1	67.9	74.1	4.2
District of Columbia	74.9	71.7	77.9	71.8	68.5	74.8	4.3
Florida	77.9	76.0	79.7	71.5	69.6	73.4	9.0
Georgia	75.1	71.8	78.2	67.6	64.8	70.3	11.1
Hawaii	79.5	76.2	82.4	73.3	69.6	76.7	8.5
Idaho	82.1	78.3	85.5	72.2	68.4	75.7	13.7
Illinois	77.8	76.1	79.5	68.3	66.0	70.6	13.9
Indiana	71.7	68.7	74.4	62.5	59.2	65.7	14.7
Iowa	74.9	72.3	77.4	67.8	64.9	70.6	10.5
Kansas	76.0	71.6	79.9	70.5	67.2	73.6	7.8
Kentucky	68.6	64.0	72.8	59.5	56.0	62.9	15.3
Louisiana	60.7	55.0	66.2	60.6	56.3	64.7	0.2
Maine	77.1	73.8	80.1	70.2	67.4	72.9	9.8
Maryland	79.3	76.7	81.7	73.4	70.8	75.8	8.0
Massachusetts	79.4	76.8	81.9	75.8	72.8	78.6	4.7
Michigan	74.5	71.7	77.2	68.0	65.4	70.5	9.6
Minnesota	82.4	80.1	84.5	72.4	69.9	74.7	13.8
Mississippi	58.4	53.6	63.0	51.8	47.4	56.1	12.7
Missouri	70.5	67.6	73.2	63.7	60.5	66.7	10.7
Montana	72.1	67.3	76.4	64.1	59.8	68.1	12.5
Nebraska	72.0	68.9	74.9	70.0	66.8	73.0	2.9
Nevada	79.5	76.3	82.4	72.4	69.1	75.4	9.8
New Hampshire	83.0	80.7	85.1	78.7	76.3	81.0	5.5
New Jersey	79.6	77.1	81.9	76.8	74.0	79.4	3.6
New Mexico	71.6	65.8	76.8	61.7	57.2	66.1	16.0
New York	75.8	73.7	77.8	70.6	68.7	72.5	7.4
North Carolina	72.9	69.8	75.8	63.9	61.0	66.6	14.1
North Dakota	79.5	76.3	82.4	67.1	63.5	70.5	18.5
Ohio	71.3	68.5	73.9	66.8	64.4	69.2	6.7

Table 5. (Continued)

State	2012 95% Conf. Interval			2006 95% Conf. Interval			Est. % Chg. 09-12
	Est.	Lower	Upper	Est.	Lower	Upper	
Oklahoma	71.2	65.2	76.5	60.6	56.9	64.2	17.5
Oregon	84.8	80.4	88.4	76.0	72.9	78.8	11.6
Pennsylvania	74.4	72.4	76.4	67.3	65.0	69.5	10.5
Rhode Island	76.5	73.5	79.2	71.6	68.6	74.3	6.8
South Carolina	69.0	66.4	71.5	58.3	54.6	61.8	18.4
South Dakota	74.0	70.3	77.3	65.5	62.2	68.6	13.0
Tennessee	68.5	63.8	72.8	62.3	58.8	65.7	10.0
Texas	69.7	67.7	71.7	63.9	62.1	65.8	9.1
Utah	82.5	77.6	86.5	77.9	74.2	81.2	5.9
Vermont	80.1	77.0	82.9	70.8	67.7	73.7	13.1
Virginia	73.8	71.2	76.3	71.0	68.2	73.7	3.9
Washington	80.2	77.5	82.7	77.9	75.1	80.5	3.0
West Virginia	63.6	59.6	67.4	59.7	55.8	63.5	6.5
Wisconsin	74.3	70.6	77.7	71.8	69.1	74.4	3.5
Wyoming	76.1	71.9	79.9	69.8	66.3	73.0	9.0

REFERENCES

Ahsan, K., Khan, O. & Salam, A. (2013). Assistive technology for night blindness: A mobile application approach. *International Journal of Mobile Learning and Organisation, 7* (2), 140-157.

Arthur D. Little, Chalmers University of Technology, & Ericsson. (2013, Sep.). *Socioeconomic effects of broadband speed.* Retrieved from http://www.ericsson.com/res/thecompany/docs/corporate-responsibility/2013/ericsson-broadband-final-071013.pdf

Cisero, M. (2014, April 17). How baby boomers are embracing mobile shopping. *BIA/Kelsey Local Media Watch.* Retrieved from http://blog.biakelsey.com/index.php/2014/04/17/how-baby-boomers-are- embracing-mobile-shopping/

DeGusta, M. (2012, May 9). Are smart phones spreading faster than any technology in human history? *MIT Technology Review.* Retrieved from http://www.technologyreview.com/news/427787/are-smart-phones-spreading- faster-than-any-technology-in-human-history/

Economics and Statistics Administration (ESA), & National Telecommunications and Information Administration (NTIA). (2000,

Oct.). *Falling through the net: Toward digital inclusion—A report on Americans' access to technology tools*. Retrieved from http://www.ntia.doc.gov/files/ntia/publications/fttn00.pdf

Executive Order No. 13636, 78 F.R. 11739 (2013).

Federal Communications Commission (FCC) & National Telecommunications and Information Administration (NTIA). (2014, Feb.). Access to broadband technology by speed. *National Broadband Map Broadband Statistics Report*. Retrieved from http://www.broadbandmap.gov/download/Technology%20by%20Speed.pdf

Federal Communications Commission (FCC) & National Telecommunications and Information Administration (NTIA). (2014, Feb.). Broadband availability in urban vs. rural areas. *National Broadband Map Broadband Statistics Report*. Retrieved from http://www.broadbandmap.gov/download/Broadband%20Availability%20in%20Rural%20vs%20Urban%20Areas.pdf

Frost & Sullivan. (2014, May 15). *Frost and Sullivan: A market potential of USD$731 billion in connected living by 2020*. Retrieved from http://www.frost.com/prod/servlet/press-release.pag?docid=290755603

Gartner. (2013, Dec. 12). *Gartner says the Internet of things installed base will grow to 26 billion units by 2020*. Retrieved from http://www.gartner.com/newsroom/id/2636073

Internet Innovation Alliance. (2013). *Access to broadband internet: Top ten areas of saving—2013*. Retrieved from http://www.internetinnovation.org/library/special-reports/access-to- broadband-internet-top-ten-areas-of-saving-2013/

Jackson, D. (2012, Jul. 30). Smartphones: The new weapon to combat unemployment. *Jobs & Hire*. Retrieved from http://www.jobsnhire.com/articles/2606/20120730/jobs-hunting-mobile-apps-iphone-android-careers-entry-level-unemployment.htm

Lowrey, A. (2014, Apr. 30). Changed life of the poor: Better off, but far behind. *The New York Times*. Retrieved from http://www.nytimes.com/2014/05/01/business/economy/changed-life-of-the- poor-squeak-by-and-buy-a- lot.html?emc=edit_th_20140501&nl=todaysheadlines&nlid=62921653&_r=0

Mandel, M. (2013, Jul. 8). 752,000 app economy jobs on the 5th anniversary of the app store. *Progressive Policy Institute*. Retrieved from http://www.progressivepolicy.org/2013/07/752000-app-economy-jobs-on-the-5th-anniversary-of-the-app-store/

Mulpuru, S. (2014, May 12). US mobile phone and tablet commerce forecast, 2013 to 2018. *Forrester Research*. Retrieved from http://www.forrester.com/US+Mobile+Phone+And+Tablet+Commerce+Forecas t+2013+To+2018/fulltext//ERES115514?intcmp=blog:forrlink

National Telecommunications and Information Administration (NTIA). (2001-2012). Current Population Survey data. *Internet and Computer Use Studies and Data Files*. Retrieved from http://www.ntia.doc.gov/data

National Telecommunications and Information Administration (NTIA). (2013, Aug. 5). *New broadband map data shows progress, but work remains*. Retrieved from http://www.ntia.doc.gov/blog/2013/new-broadband-map-data-shows-progress- work-remains

National Telecommunications and Information Administration (NTIA). (2014, Feb. 5). *Digital learning program prepares students for school, careers and life in the 21st century*. Retrieved from http://www.ntia.doc.gov/blog/2014/digital-learning-program-prepares-students-school-careers-and-life-21st-century

National Telecommunications and Information Administration (NTIA). (2014, Jun. 5). *NTIA releases interim progress report on administration's plan to free up more spectrum.* Retrieved from http://www.ntia.doc.gov/blog/2014/ntia-releases-interim-progress-report-administration-s-plan-free-more-spectrum

National Telecommunications and Information Administration (NTIA) & Economics and Statistics Administration (ESA). (2013, June). *Exploring the digital nation: America's emerging online experience*. Retrieved from http://www.ntia.doc.gov/files/ntia/publications/exploring_the_digital_nation_ -_americas_emerging_online_experience.pdf

Office of Management and Budget (OMB). 2009, Dec. 9), *OMB bulletin no. 10-02: Update of statistical area definitions and guidance on their uses*. Retrieved from http://www.whitehouse.gov/sites/default/files/omb/assets/bulletins/b10-02.pdf

The Organisation for Economic Co-operation and Development (OECD). (2014a, Jan. 9). Total fixed and wireless broadband subscriptions by country. *OECD Broadband Portal.* Retrieved from http://www.oecd.org/sti/broadband/oecdbroadbandportal.htm

The Organisation for Economic Co-operation and Development (OECD). (2014b, Jan. 9). Fixed and wireless broadband subscriptions per 100 inhabitants. *OECD Broadband Portal.* Retrieved from http://www.oecd.org/sti/broadband/oecdbroadbandportal.htm

Pew Research Center. (2014, Jan. 16). E-reading rises as device ownership jumps. *PewResearch Internet Project*. Retrieved from http://www.pewinternet.org/2014/01/16/e-reading-rises-as-device-ownership-jumps/

Rogers, E.M. (1995). *Diffusion of innovations* (4th ed.). New York, NY: Free Press.

Salpeter, M. (2012, Jul. 11). The 9 best tips for submitting an online job application. *US News & World Report*. Retrieved from http://money.usnews.com/money/blogs/outside-voices-careers/2012/07/11/the-9-best-tips-for-submitting-an-online- job-application

SQW Group. (2013, Nov.) UK broadband impact study: Impact report. *Commissioned by the United Kingdom Department for Culture, Media & Sport.* Retrieved from https://www.gov.uk/government/uploads/system/uploads/attachment_data/file/257006/UK_Broadband_Impact_Study _-_Impact_Report_-_Nov_2013_-_Final.pdf

U.S. Bureau of Labor Statistics (BLS). (2014). Labor force characteristics. *Labor force statistics from the Current Population Survey*. Retrieved from http://www.bls.gov/cps/lfcharacteristics.htm#nlf

U.S. Bureau of Labor Statistics (BLS) & U.S. Census Bureau. (2006, Oct.). Technical paper 66. *Design and Methodology—Current Population Survey,* 5-1. Retrieved from http://www.census.gov/prod/2006pubs/tp-66.pdf

U.S. Census Bureau. (2011). *July 2011 computer and Internet use supplemental survey.* Retrieved from http://thedataweb.rm.census.gov/ftp/cps_ftp.html#cpssupps

U.S. Census Bureau. (2011). *July 2011 current population survey.* Retrieved from http://thedataweb.rm.census.gov/ftp/cps_ftp.html#cpsbasic

U.S. Census Bureau. (2012). *Current population survey (CPS) data.* Retrieved from https://www.census.gov/cps/data/

U.S. Census Bureau. (2012). *October 2012 current population survey.* Retrieved from http://thedataweb.rm.census.gov/ftp/cps_ftp.html#cpsbasic

U.S. Census Bureau. (2012). *October 2012 school enrollment and Internet use supplemental survey.* Retrieved from http://thedataweb.rm.census.gov/ftp/cps_ftp.html#cpssupps

U.S. Department of Commerce. (2013, July 31). *Commerce's Internet Policy Task Force releases report on digital copyright policy.* Retrieved from http://www.commerce.gov/blog/2013/07/31/commerces-internet-policy-task- force-releases-report-digital-copyright-policy

The White House, Office of the Press Secretary. (2013, June 6). *President Obama unveils connected initiative to bring America's students into digital age.* Retrieved from http://www.whitehouse.gov/the-press-office/2013/ 06/06/president-obama-unveils-connected-initiative-bring-america-s-students-di

The White House, Office of the Press Secretary. (2013, June 6). *Remarks by the President at Mooresville Middle School.* Speech presented at Mooresville Middle School, Mooresville, NC. Retrieved from http://www.whitehouse.gov/the-press- office/2013/06/06/remarks-president-mooresville-middle-school-mooresville- nc

End Notes

[1] Two studies issued in 2013 support the notion that broadband has salient effects. One study asserts that based on an 11-country analysis, the faster the connection speed, the greater the quantifiable benefits to household income and ultimately a nation's gross domestic product (Arthur D. Little, Chalmers University of Technology, & Ericsson, 2013); while the other argues that for every £1 the government invests in broadband, the UK economy benefits by £20, and that social and environmental benefits also result from broadband use (SQW Group, 2013).

[2] Among countries that are members of the Organisation for Economic Co-operation and Development (OECD), a global organization of nations with market economies backed by democratic institutions, the United States ranked number one in terms of total wired (91 million) and wireless (299 million) broadband subscriptions as of June 2013. Arrayed by subscriptions per 100 inhabitants, the United States ranked 16th in wired connections and 7th in wireless (OECD, 2014a & 2014b).

[3] The CPS public use files do not provide the geographic variable for identifying a household's location as urban or rural. In this report, "urban" refers to metropolitan areas and "rural" to nonmetropolitan areas. Since 2000, a metropolitan area is defined by its "core based statistical area" (CBSA), which includes both metropolitan and micropolitan statistical areas. According to the 2000 standards, each CBSA must have at least one urban area with at least 10,000 inhabitants. Each metropolitan statistical area must contain at least one urbanized area with a population of 50,000 or more. Each micropolitan statistical area must contain at least one urban cluster with a population of between 10,000 and 50,000 (Office of Management and Budget [OMB], 2009).

[4] A "householder" or "reference person" is the household member who owns or rents the housing unit (U.S. Bureau of Labor Statistics & U.S. Census Bureau, 2006). When analyzing household-level data, NTIA uses the householder's demographic characteristics (such as age, race, and educational attainment) as a proxy for the entire household.

[5] It is important to note that, due to the smaller sample sizes for individual states, state-level adoption estimates are subject to a margin of error of several percentage points. These estimates therefore are best understood as a general assessment of home Internet use in each state, and not as a precise ranking.

[6] It is beyond the scope of this study to attempt to identify the causes of different growth rates observed in each state. In some states with high adoption rates, for example, saturation may simply have limited growth in home Internet use. Other states may have faced economic issues, population shifts, or a range of other factors. NTIA encourages researchers to delve more deeply into the circumstances that affect growth in Internet adoption.

[7] Frost & Sullivan project the market for "connected living" or ubiquitously connected video and data services at home, work, and public spaces will grow to $731.7 billion by 2020, citing cloud computing, big data, mobility, and low cost sensors as driving growth of the Internet of Things (IoT). Fifteen percent of this estimate ($111 billion) is attributable to the "connected home" (2014). Further, Gartner predicts that the low cost of processors will cause the number of IoT devices, which communicate or interact with external environments or internal systems, to grow 3.5 times larger than the 7.3 billion smartphones, tablets, and personal computers expected in use by 2020. Also, the IoT will yield $1.9 trillion in global economic valued added through sales in diverse end markets (2014).

[8] The U.S. Department of Labor defines the term as "[p]ersons who are neither employed nor unemployed are not in the labor force. This category includes retired persons, students, those taking care of children or other family members, and others who are neither working nor seeking work. Information is collected on their desire for and availability for work, job search activity in the prior year, and reasons for not currently searching." (U.S. Bureau of Labor Statistics [BLS], 2014).

[9] The falling prices of certain consumer goods and services such as personal computers, cell phones, and cell phone service make them affordable for low income consumers, while "the cost of many services crucial to escaping poverty—including education, health care, and child care—has soared" (Lowrey, 2014).

[10] The specific questions used in the October 2012 survey instrument are available at http://www.census.gov/prod/techdoc/cps/cpsoct12.pdf.

In: America's Embrace of the Mobile Internet ISBN: 978-1-63463-585-1
Editor: Jennifer Haggard © 2015 Nova Science Publishers, Inc.

Chapter 2

THE IMPACT OF CULTURE ON THE ADOPTION, DESIGN AND USAGE OF MOBILE DEVICES*

Susan G. Hill

1. INTRODUCTION

The use of mobile devices for communications and information exchange has been shown to be very desirable, with more than 5.3 billion mobile subscribers estimated (mobiThinking, 2011) and more than half-billion mobile phones sold each year (Blom, Chipchase, and Lehikoinen, 2005). Most often, when the term "mobile devices" is used, the authors really mean "mobile phones." For example, Blom et al. (2005) describes "personal, mobile synchronous and asynchronous communication for all types of users..." (p. 37) and then cites that more than one half a billion mobile phones are sold each year. However, mobile devices are not limited to phones; in addition to cellular phones, there are also personal digital assistants (e.g., Blackberry) as well as the more recent advent of wireless tablet computers (e.g., iPad). With these "smart" mobile devices, users can communicate by voice and text as well as have Internet access as a means of communications and information exchange. Mobile device use has grown tremendously; for example, in 2002,

* This is an edited, reformatted and augmented version of a report, ARL-TR-6183, issued by the Army Research Laboratory, October 2012.

mobile (cellular) phones exceeded the number of landlines (Donner, 2008). All types of mobile devices have increased as well; for example, it is estimated that 25 million iPads have been sold to date (Rosoff, 2011).

The purpose of this report is to explore some of the research literature that discusses mobile technology, specifically mobile phones and mobile access to the Internet, and the impact that culture has on its adoption, use and design. One of the results of this review will be to identify factors that are important when considering mobile technology and culture, specifically, and factors that may be important in examining the impact of culture on decision making and communication, more generally. To accomplish this goal, the report first looks at culture, its definition, models of culture and dimensions used to describe specific cultures. The next section then examines mobile technology and culture, focusing on (1) models of adoption of technology, specifically mobile technology; (2) examples of uses of mobile technology and the influence of culture, and (3) design of mobile user interfaces and experience considering cultural influences. Finally, we discuss the information identified and its relevance to our goals of building a cognitive framework for social-cultural modeling, as well as other observations concerning technology and culture, especially military applications.

2. Dimensions and Models of Culture

There are a number of definitions of culture. Aryana and Oritsland (2010) say that, "In 1952 Kroeber and Kluckholm reviewed different definitions of culture and found 162 definitions. Now there are even more..." (p. 217). An oft quoted definition is that of Hofstede (1980), which defines culture as "the collective programming of the mind that distinguishes the members of one group or category of people from another" (quoted, for example, in Min, Li and Ji. 2009). Van Biljon (2006) proposes a definition of culture specific to mobile devices: "The patterns of thinking, feeling and acting that influence the way in which people communicate among themselves and use mobile devices." (p. 34)

Researchers have tried to define culture in terms of objective, measurable elements. The most well known of culture dimensions are those proposed by Hofstede (e.g., Hofstede and McCrae, 2004). There are a number of other attempts made to describe cultural dimensions: Min, Li and Ji (2009) identified a number of these cultural dimensions commonly cited and combined them into a table of cultural dimensions, presented in table 1.

Table 1. Cultural Dimensions (adapted from Min, Li and Ji, 2009)

Cultural Dimension	Description	Authors
Individualism/Collectivism	Individualism is a primary orientation to the self; Collectivism a primary orientation to common goals and objectives.	Hofstede (1980), Kluckhohn and Strodtbeck (1961), Trompenaars (1993)
Uncertainty avoidance	The extent to which members of a culture feel threatened by uncertain or unknown situations.	Hofstede (1980),
Contextuality, Context	High-context cultures prefer implicit messages and indirect communication; low-context cultures explicit messages and direct communication.	Hall (1976)
Power distance	The extent to which members of a culture accept unequal distributions of power.	Hofstede (1980),
Masculinity/femininity	The degree to which a society is characterized by assertiveness (masculinity) or nurturances (femininity).	Hofstede (1980),
Notions of time	Monochromic and polychromic; past, present, and future orientation; short- and long-term orientation.	Hall (1976), Trompenaars (1993)
Affectivity/neutrality	The range of feeling and emotions expressed in public.	Parsons and Shils (1951), Trompenaars (1993)
Universalism/particularism	The manner in which members of a culture adhere to agreed standards; Universalism is rules based, particularism is relationship based.	Parsons and Shils (1951), Trompenaars (1993)

Table 1. (Continued)

Cultural Dimension	Description	Authors
Ascription/achievement	Ascription gives priority to certain attributes individuals possess, whereas achievement gives priority to their accomplishments over their given attributes.	Kluckhohn and Strodtbeck (1961), Parsons and Shils (1951), Trompenaars (1993)
Specificity/diffuseness	The degree to which individuals become involved with one another.	Parsons and Shils (1951), Trompenaars (1993)
Human relation to natural and social world	The extent to which the human relation to the natural and social world is understood as one of mastery or one of harmony.	Kluckhohn and Strodtbeck (1961), Schwartz and Ros (1995), Trompenaars (1993)

Van Biljon (2006) and van Biljon and Kotze (2007a, b; 2008) also have extensive discussions on a number of theoretical issues and models of culture. In addition to Hofstede, they also mention the onion model of Trompenaars (1993), Hall (1976), and others. Bagchi and Kirs (2009) examined Schwartz's cultural values (Schwartz, 1992) as applied to information and communication technologies. These authors specifically look at personal computer (PC) and Internet use and do not address mobile devices. While other cultural studies discuss Schwartz's cultural values, no studies were found that apply those cultural dimensions to mobile devices; therefore, we can make no assessment as to the usefulness of that cultural framework to mobile device adoption, design and usage.

Van Biljon and Kotze (2007a) discuss objective vs. subjective culture. Objective culture is related to concrete things that are "visible, tangible and easy to examine" (p. 2), for example, language, currency, date and time formats, and color preferences. Subjective culture is related to psychological and sociological things such as humans' assumptions, beliefs, values, attitudes and patterns of thinking. Designing for the objective culture is largely understood, although some studies suggest otherwise (van Biljon and Kotze, 2007a). Subjective culture does not seem to be as well understood and, therefore, is more difficult to describe, assess and to determine the implications for technology design and human behavior.

Baumgartner (2003) described and ranked 50 cultural dimensions related to human-computer interaction. Three of the first ten dimensions are drawn from Hofstede (1994): uncertainty avoidance (no. 3), power distance (no. 5), and individualism/collectivism (no. 10.) Hofstede's time (1994) orientation dimension was rated as no. 11 and his masculinity vs. femininity dimension rated at number 14. Van Biljon and Kotze (2007b, p. 4) used the descriptions from Baumgartner (2003) to define the other seven (non-Hofstede) dimensions. Table 2 shows the Baumgartner dimensions and the related Hofstede (1994) dimensions.

Table 2. The top cultural dimensions as identified by Baumgartner (2003) and the relationship to the five dimensions of Hofstede (1994)

Baumgartner (2003) Grouped and Ranked Cultural Dimensions and Definitions (p. 48)	Hofstede (1994)
1. Context *Context*, referring to the amount and specificity of information in a given situation, and scaling from *high context* communication, which has most of the meaning in the context, to *low context* where the complete meaning is in the message.	—
2. Technological Development *Technological development*, referring to the rate of technological development, and scaling from *advancement* to *backwardness*.	—
3. Uncertainty Avoidance *Uncertainty avoidance* is the extent (high or low) to which the members of a culture feel threatened by uncertain or unknown situations. Cultures vary in their avoidance of uncertainty, creating different rituals and having different values regarding formality, punctuality, legal-religious-social requirements, and tolerance for ambiguity.	Uncertainty Avoidance
4. Time perception *Time perception*, where *monochronic* time perception is characterized by schedules, promptness and compartmentalization of activities. The other end of the scale represents *polychonic* time perception which is characterised by people doing several things at a time. Schedules may be changed and time commitments are desirable rather than absolute.	—

Table 2. (Continued)

Baumgartner (2003) Grouped and Ranked Cultural Dimensions and Definitions (p. 48)	Hofstede (1994)
5. Authority conception; Power distance *Authority conception* or *Power distance* is the degree to which people favor an authoritarian, egalitarian, paternalistic, or participative management style refers to the extent (high or low) to which less powerful members of institutions and organizations within a country expect and accept that *power is distributed unequally.*	Power distance
6. Affective vs. neutral *Affective vs. neutral*, refer to the way humans express their emotions, and scaling from *high affective (*expressing feelings openly) to *low affective.*	—
7. Face-saving; Specific vs. diffuse; Instrumental vs. expressive *Face-saving*, referring to the act of reserving one's prestige or outward dignity, and scaling from *high face saving to low face saving.*	—
8. Activity orientation; Meaning of life *Activity orientation*, referring to the extent to which activity is valued within a culture. Human activity can focus on *doing*, *being* or *beingin-becoming.*	—
9. Nonverbal communication; Space *Nonverbal communication*, referring to differences in behaviour as expresses in: "k*inesics* (body movement and facial gestures); *proxemics* (distance); *oculesics* (eye movements and eye contact); *haptics* (touching behavior); *paralanguage* (tone of voice and non-language sounds); and *appearance* (dress and grooming)".	—
10. Individualism vs. collectivism *Individualism vs. collectivism* refers to the role of the individual and group, and which interest prevails over the other. For example, in individual-oriented countries everyone is expected to look after one's self or immediate family. Collectivism implies that people are integrated from birth into strong, cohesive groups that protect them in exchange for unquestioning loyalty.	Individualism vs. collectivism

Baumgartner (2003) Grouped and Ranked Cultural Dimensions and Definitions (p. 48)	Hofstede (1994)
11. Time orientation; Long-term vs. short-term orientation *Time orientation* refers to the way cultures conform to time. *Past Orientation* implies a belief that understanding the past can guide current and future experience. *Present Orientation* implies dominance of the mental state of the moment. *Future Orientation* implies expectancy of advancement, improvement or progression. Longterm orientation means looking for meaning and results over long periods of time (similar to past or future orientation) as opposed to viewing meaning and results over a very short period of time (similar to present orientation).	Time orientation
12. Universalism vs. particularism *Universalism vs. particularism* describes the degree of adhering to agreed standards. *Universalists* believe that good and right are definite and always applicable. *Particularists* place more emphasis on the unique circumstances.	—
13. International trade and communication *International trade and communications* refers to the rate of development in the field of trade and communication with other countries/cultures (for example, isolation vs. cooperation).	—
14. Gender roles *Gender roles* portrays the extent within cultures to which social gender roles are clearly distinct. For example, in highly distinct (for example, *masculine* or assertive) cultures men are supposed to be assertive, tough and focused on material success, whereas women are supposed to be more modest, tender and concerned with the quality of life. Low or no distinction (that is, *feminine* or modest) cultures pertains to societies in which social gender roles overlap. Both men and women are supposed to be modest, tender, and concerned with the quality of life.	Masculinity vs. femininity

The culture dimensions that researchers have derived can be useful in trying to categorize and quantify aspects of culture. These cultural dimensions will be referred to in many of the studies reviewed. However, it is beyond the scope of this report to present an in-depth analysis of cultural models and dimensions and the strengths and weaknesses of them.

Baumgartner (2003) proposes a means for visualizing and comparing the cultural dimensions of one or more countries based on a shape diagram using the number of relevant cultural dimensions. In the example, a pentagon is used for five cultural dimensions. Given that there is a quantitative measure of each cultural dimension, the dimension value can be plotted on the pentagon. In the example shown in figure 1, the two countries are similar in values for Time Perception, but are most different in Context. Baumgartner presents this approach as a means to see where localization efforts for human-computer interaction and interface design should be addressed. This approach could be used for any comparison of cultural dimensions, like for mobile device interfaces, if quantitative values are available.

Country 1 ▢		Country 2 ■	
Context	100	Context	1
Technological Development	34	Technological Development	56
Uncertainty Avoidance	54	Uncertainty Avoidance	37
Time Perception	27	Time Perception	28
Authority Conception	38	Authority Conception	90

(a)

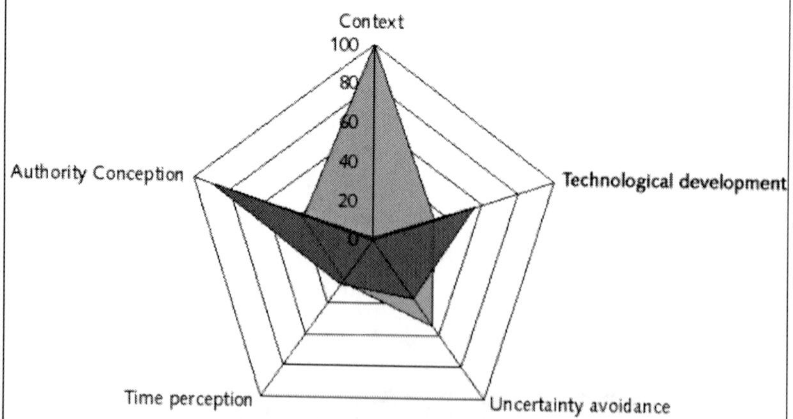

(b)

Figure 1. Using a cultural pentagon as a means to visualize and compare cultural dimensions. The quantitative measures from countries 1 and 2 are shown in (a) above and then visualized in the pentagon below (b) (Baumgartner, 2003).

3. CULTURE AND MOBILE DEVICES

There is a large and growing literature specifically addressing issues related to culture and mobile devices. An initial grouping of the research issues suggest three categories: (1) issues dealing with the adoption of mobile technology in various cultures, (2) the impact of culture on the technology usage, and (3) the design of mobile systems, specifically applying human-computer interaction and human factors knowledge related to the design of the user interface and user experience. This is similar to the approach used by Donner (2008). He reviewed ~200 recent studies concerning mobile (cell) telephones in the developing world and identified common themes in the literature. He then categorized the studies along various dimensions, including adoption, impact and the interrelationship between technologies and users. Each of these areas will be discussed.

3.1. Adoption of Mobile Technology and Culture

With any new technology, individuals must decide to use the technology before it can be helpful for a number of different applications. Adoption is the decision to use a technology. This is true, for example, for commerce conducted via mobile device, often called m-commerce. A broad range of users must adopt (i.e., decide to use) mobile technology before m-commerce will be widespread, therefore, those interested in promoting m-commerce are particularly interested in better understanding factors that affect adoption of mobile communications technology.

There are a number of models of technology adoption. One model is called the Technology Acceptance Model (TAM), which says that a user's adoption decision is based on perceived usefulness and perceived ease of use (Davis, Bagozzi, and Warshaw, 1989). The TAM is shown in figure 2. As the name implies, this focuses primarily on technology acceptance and use. A more recent, updated model of technology acceptance is called the Unified Theory of Acceptance and Use of Technology (UTAUT) (Venkatesh, Morris, Davis, and Davis, 2003). The UTAUT, shown in figure 3, is intended to consolidate key concepts from previous models. The concepts of performance expectancy, effort expectancy, social influence, and facilitating conditions are mediated by gender, age, experience, and voluntariness to determine usage intention and behavior. The definitions of these mediated concepts, as given by the authors, are:

- Performance expectancy is defined as “the degree to which an individual believes that using the system will help him or her to attain gains in job performance.” (p. 447). This is the strongest predictor of the intention to use.
- Effort expectancy is defined as “the degree of ease associated with the use of the system” (p. 450). This is important in initial usage, but lessens in importance as there is more usage.
- Social influence is defined as “the degree to which an individual perceives that important others believe he or she should use the new system” (p. 451). This is more important when usage is mandated and less important when usage is voluntary.
- Facilitating conditions are defined as “the degree to which an individual believes that an organizational and technical infrastructure exists to support use of the system” (p. 453).

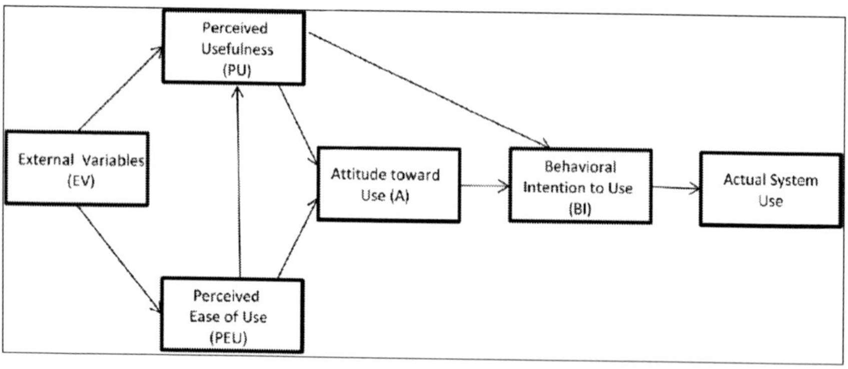

Figure 2. TAM (from Davis et al., 1989).

Min, Li, and Ji (2009) suggest that m-commerce adoption requires more than just technology acceptance; it also entails the user to become part of a network, particularly a social network. Users also have roles as consumers in m-commerce applications. Therefore, adopting m-commerce is more than just adopting technology, it is also adopting roles of consumer and network member. Network member adoption refers to subjective norms influenced by a network of others, not just based on individual decisions. M-commerce users are primarily customers using businesses and may be more interested in things that benefit themselves, such as cost, trust, and enjoyment. In this case, consumers’ adoption decisions include perceived enjoyment and personal innovativeness.

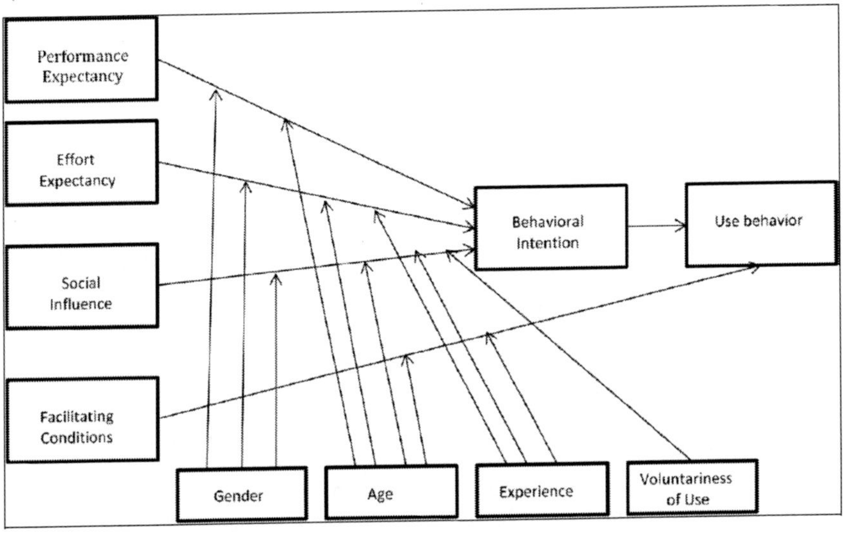

Figure 3. Diagrammatic representation of the UTAUT (from Venkatesh et al., 2003.).

The authors were interested in examining how individual-level cultural values impact the adoption of m-commerce services. They particularly framed the examination in terms of each user having multiple roles of technology user, network member, and consumer. The researchers found that values of individual culture do impact adoption decisions. For example, individuals with high uncertainty avoidance (UA) look to their network membership (i.e., their social influence) for information and cues that m-commerce services should be adopted (i.e., used by the individual.) Therefore, to promote adoption, emphasis on network memberships could be used. For example, utilizing some form of "friends and family" approach to promoting m-commerce would be thought to have impact, especially for those with high uncertainty avoidance. Similarly, those with high collectivistic cultural values would be receptive to an emphasis on collaborative services such as instant messaging rather than just stand-alone services. The authors' perspective of multiple roles for each user, including technology user, network member, and consumer, provides a better understanding of the interplay between roles, decisions to use technology, and the impact of cultural values. Based on this background and these ideas, Min, Li, and Ji propose an m-commerce adoption model based on individual level cultural characteristics, shown in figure 4.

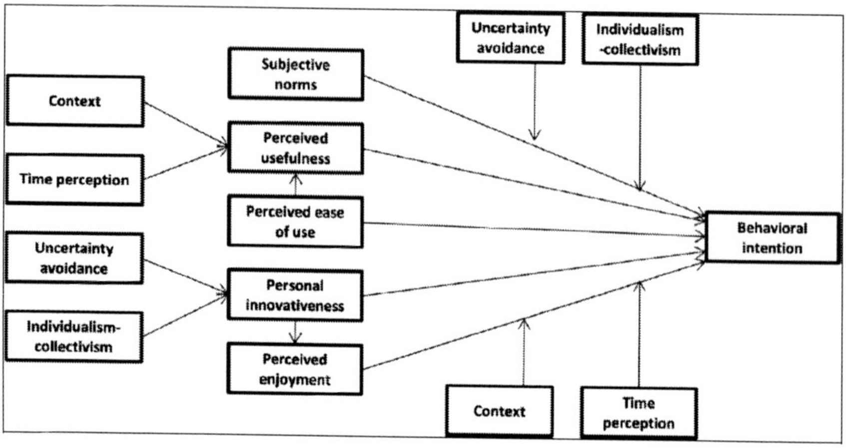

Figure 4. A model of m-commerce adoption based on individual-level cultural characteristics. (from Min, Li, and Ji, 2009).

Definitions of the terms used in the model shown in figure 4 are derived from those discussed in Min, Li and Ji (2009). These definitions are:

- Uncertainty avoidance – the degree to which people within a culture feel anxious or threatened by uncertain or unknown situations.
- Individualism-collectivism – individualism has a main orientation towards the self; collectivism has a main orientation towards the group and common goals and objectives.
- Context – context deals with the amount of information conveyed within a communication. High context cultures expect implicit information and indirect communication; low context cultures expect explicit messages and direct communication.
- Time perception – how cultures perceived time and the degree to which cultures are oriented to the past, present, and future; or short-term or long-term orientation.
- Subjective norms – how a person thinks that most people who are important to him would view him if he performs or does not perform the behavior in question.
- Behavioral intention – the intention to exhibit a particular behavior, in this case intent to adopt the use of m-commerce.
- Perceived usefulness – how useful the adopted m-commerce technology is thought to be.

- Perceived ease of use – how easy the adopted m-commerce technology is perceived to be.
- Personal innovativeness – the willingness of a person to try out new technology.
- Perceived enjoyment – the extent to which the system is fun to use.

Kwon and Chidambaram (2000) proposed a model of cell phone use that includes consideration of culture. Culture is embedded in the demographic factor of nationality. They studied cell phone adoption and use in an urban setting. They developed and tested a research model of cell phone use, based on earlier models of technology acceptance such as TAM, and tested the model in an urban area in Hawaii. They found the users were influenced directly or indirectly by: (1) individual characteristics; (2) perceived ease of use; (3) perceived usefulness (i.e., extrinsic motivations); (4) enjoyment/fun (i.e., intrinsic motivations); and (5) social pressure. Apprehensiveness about technology was also included in the model. These factors, then, influenced the individual's decision to adopt and use cell phones. The model is shown in figure 5.

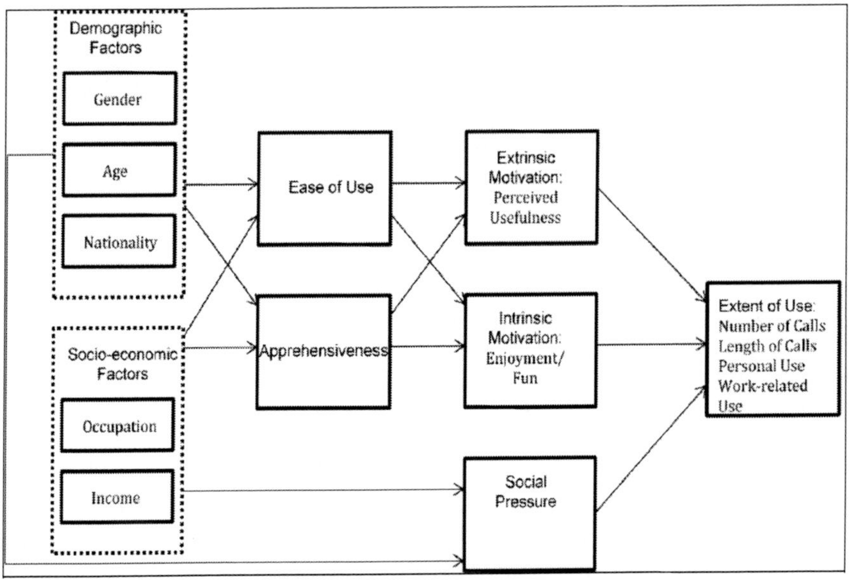

Figure 5. Research model on cell phone adoption from Kwon and Chidambaram, 2000.

Sarker and Wells (2003) present an input-process-output model of mobile device adoption and use. They say that instead of using well-established models, such as TAM, that consider technology in general, they wanted to explore the important and unique factors that specifically contribute to mobile device use. Therefore, the model proposed in their paper addresses mobile handheld devices specifically and identifies factors unique to mobile devices, such as types of user mobility, network considerations, and characteristics of the network services available. The model is shown in figure 6. In the model, culture is considered within the individual characteristics that influence the decision to adopt and use mobile devices.

Van Biljon and Kotze (2008) also address the importance of cultural factors in adoption and usage of mobile phones. After review of research on culture, mobile phones and technology adoption, they integrated the factors that influence mobile phone adoption and usage. There are determining factors and mediating factors, similar to the two types of factors in UTAUT.

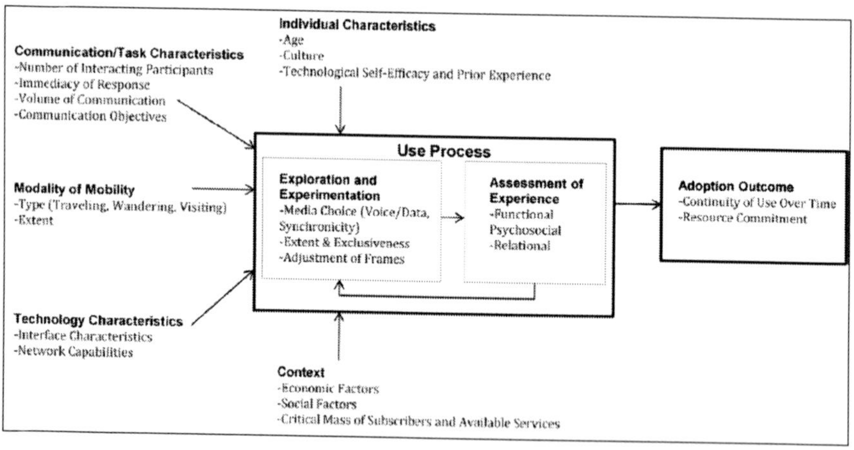

Figure 6. An Input-Process-Output model of mobile device use and adoption (from Sarker and Wells, 2003).

Van Biljon and Kotze (2007a; 2008) present the proposed model, shown in green (lighter color) in figure 7. The authors also discuss revisions that result in the final model. The final model entitled, the final Mobile Phone Technology Adoption Model (MOPTAM), is presented in figure 7 in the black (darker color). Figure 7, shows both the proposed and final models presented in green (light) and black (dark), respectively,

There are three differences between the proposed model (van Biljon and Kotze, 2008) and the final model (van Biljon and Kotze, 2007a). In the final model, facilitating conditions are seen to influence perceived usefulness. Social influence does not directly influence behavioral influence, but only has indirect (one-step removed) influence. Finally, human nature influence and cultural influence are not specifically called out in the final model. In the final model, specific influences are left undefined. However, social influence remains, contributing to both perceived usefulness and perceived ease of use.

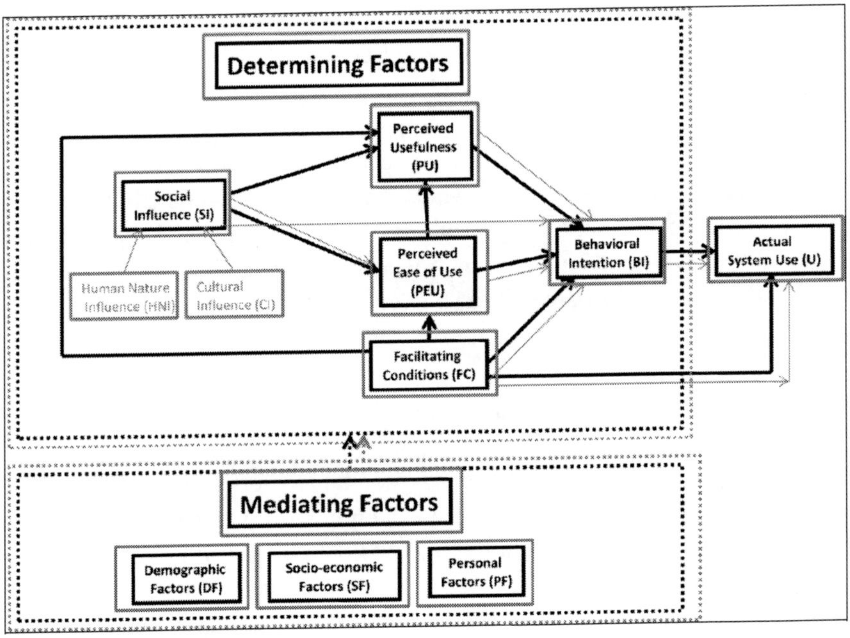

Figure 7. Proposed (van Biljon and Kotze, 2008) and final models (van Biljon and Kotze, 2007a). MOPTAM. The final MOPTAM is in black (dark) and the proposed model elements are in green (light).

The changes implemented in the proposed versus the final MOPTAM models are based on survey data collected from 59 participants, who were relatively homogenous in age (95% were between 18–30 years of age). The importance of facilitating conditions, such as phone cost, system cost, system service and system quality, in the decision to adopt mobile phone technology is reflected in the final version, with facilitating conditions influencing both perceived usefulness and actual system use. The facilitating conditions and personal factors were found to be particularly important for mobile phones as compared with other kinds of technology (van Biljon and Kotze, 2007a). Cultural influences were specifically identified in the proposal model as contributing to social influence. Note that in the final (revised) model, cultural factors are not specifically called out. Therefore, the role that culture plays in this model is left undefined. Cultural factors clearly could be placed within the mediating factors and can be associated with demographic (national-level culture) or personal factors (individual-level culture). Cultural influences could also be considered as a contributor to social influence (SI), as shown in the determining factors. While the final model does not clarify the role of culture and how and where it influences system use, culture should not be thought of as unimportant. Culture should be considered an important factor and its role should be clarified.

Mobile phones encourage small power distance, individualism and self-expression. However, mobile phones are accepted in cultures with large power distances and collectivist values, which suggest that mobile phone culture takes precedence over ethnic and cultural values and social behaviors (van Biljon, 2006). The implications of mobile technology and how it might change social and cultural norms are still to be worked out.

3.2. Mobile Technology Usage and Culture

In the previous section, some of the models refer to mobile device adoption and usage. Clearly, first there is adoption of the technology and then it is used for some purpose. Choi et al. (2005) suggest that as mobile device and services use has increased globally, cultural factors have had a more pronounced impact on mobile use than on traditional stationary Internet users. This is because wireless networks operate only in local areas whereas stationary Internet devices are "globally uniform." It is recognized that mobile commerce cannot meet its potential without widespread adoption of mobile technology (Sarker and Wells, 2003), hence the focus on technology adoption in the previous section. However, in this section, we will identify some of the uses of the mobile technology once it has been adopted and how the uses are related to culture. Mobile technology is seen as enabling *mobile communication*, *mobile collaboration*, and *mobile commerce* (Sarker and Wells, 2003)—three important areas of usage.

3.2.1. Mobile Communication

Mobile technology allows communication between people at *anytime* and *anywhere*. Communication can take place verbally (speaking on the phone) or visually (text messages or e-mail). The purpose of communications can be thought of as (1) conveyance, where a person conveys a piece of information; (2) convergence, where people converge on a shared understanding; and (3) passive reception, where people just receive information. Mobile communication is, perhaps, the most obvious use of mobile technology.

3.2.2. Mobile Collaboration

Mobile collaboration, using mobile technology, is focused on processes and tools that allow users to work together anywhere (Guerrero et al., 2006). This application is often considered in the realm of computer-supported cooperative work (CSCW). It is beyond the scope of this paper to fully review the CSCW literature. However, it is not difficult to see that culture will play an important role in the ability of distributed individuals, possibly working at different times (i.e., asynchronously), to successfully collaborate on work using mobile technology. Herskovic, Ochoa, Pino, and Neyem (2011) provide a general framework for attributes to be considered during concept development and design of mobile collaborative systems.

3.2.3. Mobile Commerce (m-Commerce)

A major reason for interest in culture and information and communication technologies is because of the possible business opportunities. Aryana and Boks (2010) point out that "new opportunities outside mature markets of developed countries may lead to more attention to user requirements in those new markets" (p. 2). The authors also point out that demographic changes, such as increasing life expectancy, can create "new consumer groups with different needs" (p. 139). If a company can make a product that is preferred over others, it can be an economic advantage in marketing and sales. M-commerce can be described as "the emerging arena within which commercial transactions are made possible using handheld devices that are connected by wireless networks." (Mathew, Sarker and Varshney, 2004, p. 1). While traditional e-commerce conducted on the Internet via desktop allows people to conduct commerce at *anytime*, mobile devices allow m-commerce to be conducted *anywhere*.

Min, Ji, and Qu (2008) discuss some fundamental differences between mobile commerce (m-commerce) and Internet-based electronic commerce (e-commerce). Some of these differences include network infrastructure, end user devices, tasks, applications, use environments, and costs and sources of revenue. Based on research conducted on user acceptance in China, the authors propose a revised UTAUT model, which would additionally include factors unique to m-commerce, Chinese culture, and user satisfaction. The additional factors are trust, privacy protection, and cost. System quality and information quality should be added as system-specific factors and culture and additional demographics should be added as additional mediating factors.

In addition to Min, Ji and Qu (2008), other researchers have considered and researched additional factors to enhance the TAM. Enhancement factors considered by other researchers that will influence the decision to adopt technology include social influences and personality traits (such as personal innovativeness), (Lu et al., 2005), perceived risk, cost and compatibility (Wu and Wang, 2005) and individual characteristics, including innovativeness, past adoption decisions, and knowledge about technology (Yang, 2005). Others factors include technology readiness and wireless website interface usability (Massey et al., 2005) as well as development of an adoption model that is value-based, including benefits and sacrifices experienced as a result of the technology adoption. All of these additional TAM enhancement factors are mentioned in a discussion of mobile services' adoption decisons by Constantiou et al. (2009).

While revisions and enhancements to the TAM model can add to our understanding of adoption, additional factors can also be important in considerations of communication, decision making, and the influence of culture on mobile technology.

There are a number of application areas where commerce is a major driver. One such area is mobile banking services. For example, Koo and Wati (2010) state that mobile banking services are considered as one of the most value-added and important mobile services available. Because of that importance, the authors looked at the role of trust as a mediating variable in mobile banking in the developing country of Indonesia. They propose a research model for use to generate hypotheses to guide research (see figure 8). They collected data from over 250 participants via questionnaires and found that their proposed model was useful for explanation of the data: trust did mediate the effects of information quality for perceived usefulness and user satisfaction and confirmed the importance of trust in mobile banking applications.

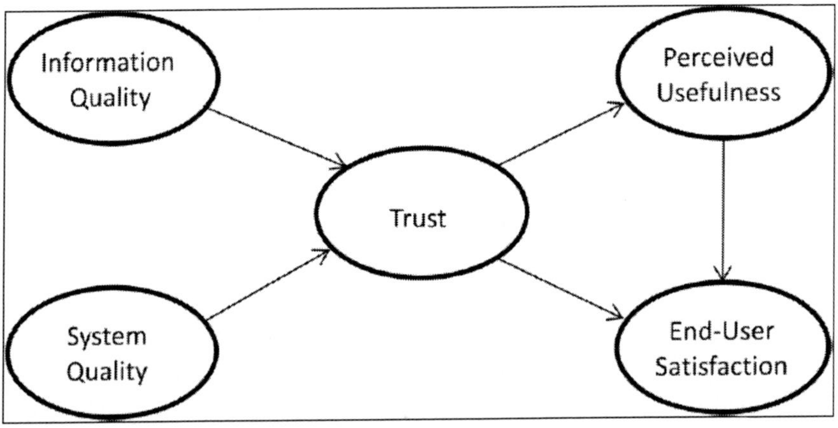

Figure 8. Research model to integrate information system success and trust. The model was used to study trust as a mediating factor among information system success factors, using mobile banking as an example (adapted from Koo and Wati, 2010).

3.2.4. Mobile Learning

Another use of mobile technology is for learning. Mobile learning (mLearning) can be considered as a meeting of mobile computing and electronic learning that results in an "anytime, anywhere learning experience." (van Biljon and Kotze, 2007b, p. 1). The authors note that context, location, and culture are important in learning in general, and may be more important for learning with mobile devices given the small screens and limitations. Further, they believe that objective culture, such as language, is an implicit part of a model of the learner but that subjective culture needs to be investigated further as a means to develop learner profiles that can be used for mLearning. (See section 2 of this report for definitions of objective and subjective culture.)

3.2.5. Other Mobile Phone Usage Issues

Several other mobile phone uses were identified that relate directly to the consideration of culture. Three areas are discussed in this section: (1) use in developing countries; (2) use for local crises; and (3) patterns of usage.

Mobile Phones in Developing Countries

An area where mobile devices, specifically mobile phones, will play an increasing important role is in developing countries. According to sources cited by Medhi et al. (2011), there were more than 4 billion phone users in 2008, and 60% of all subscribers (or 2.4 billion people) live in developing countries. So, researchers interested in global development see the mobile phone as a potential means to support development in areas such as education, finance, health, agriculture, and governance (Donner, 2008).

A major challenge within developing countries is to understand the local population and culture and design mobile technology for their use. For example, one challenge is that much of the population of developing countries is non-literate, or if literate, the individuals are novice users of computer technology and mobile user interfaces must be designed for use by novices (Medhi et al., 2011).

An example of using mobile phones for development is given in Martin and Abbott (2010), which describes the use of mobile phones in agricultural development. They conducted a study by interviewing small- to medium-size farm holders in Uganda to collect baseline data on the use, diffusion and impact of mobile phone use, specifically for agriculture-based use. The authors report five themes identified from the data for agricultural-based mobile phone use: (1) coordinating meetings, training, and access to agricultural inputs, (2) consulting with experts, (3) obtaining market information, (4) receiving agriculture emergency security information, such as livestock or crop emergencies, and (5) monitoring of financial transactions such as loans. Using a mobile phone provides a real opportunity to communicate with others without travel. Some of the unique local uses include using the calculator, using the loudspeaker function for group meetings, storage of information in the calendar, voice-recording of lessons, and use of the camera for demonstrations. The themes identified in this research directly relate to mobile communication (themes 2 and 4), collaboration (theme 1), and m-commerce (themes 3 and 5) and shows the broad range of activities that have been enabled by mobile technology that were not, or not easily, available previously. One can easily see how these potential uses will influence the decision to adopt mobile technology.

Kamppuri (2011) has developed a new cross-cultural human-computer interaction (HCI) design approach that emphasizes worth, value and user experience as major topics. This new approach is used in an example case study of mobile phone use by Tanzanian university students. Some of the mobile phone uses discussed in this case study include immediate communication that was previously unavailable because of few land lines, as well as the slowness and unreliability of mail service; staying connected to the social network (even if connections are very brief); managing group studies and assignments; and, demonstrating status. An interesting point is the distinction between ownership and access to mobile phone usage; individuals do not need to own a mobile phone in order to have access for use. Kamppuri notes that previous research states that perceived access among Tanzanian users was reported as high as 97% in 2005; sharing is considered typical (although some problems are noted.) Beeping, that is calling and hanging up quickly as a means to let the called party know that you wish them to call you back, is common, as is transferring airtime from one phone to another. These are different ways of using the phone as compared to other cultures. In addition to calling, texting, and beeping, other frequently used features include the alarm clock, calendar, calculator, radio and flashlight. This shows uses of mobile technology beyond communications. Kamppuri also notes three user characteristics that impact phone usage: gender, age, and living in urban vs. rural areas.

Mobile Phones As Devices for Local Crises

Another interesting use of mobile technology is discussed by Evers et al. (2010). They present ongoing research that describes communications by users and environmental agencies via environmental-monitoring autonomous mobile agents. The concept is that hazardous situations can be identified in populated areas by using inputs from people via mobile devices and through a distributed sensor network. For example, users can communicate with a virtual agent through the mobile phone about unusual smells. If there is an emergency, users can receive instructions from the agent. This dialogue agent can adapt interaction and communication to the cultural background of the user, including the tasks of interruption, requesting information, and instruction. It is relevant for users to perceive the agent as part of the culture because it has been shown that people are more likely to comply with agents that are seen as part of the same group (Evers et al., 2008) and more likely to change their decision to that of the agent if the agent communicates in a culturally appropriate (i.e., socially normative) way (Wang et al., 2009.)

Aloudat and Michael (2011) present a case study describing considerations in developing a mobile phone-based emergency warning system in Australia. They describe their methodology of interviews with the general public and then with key individual experts (some independent and some from the government.) The authors identified, via the interviews, many factors that the public was concerned about, using the framework of factors associated with technology acceptance, such as behavioral intent, perceived usefulness, perceived ease of use, but also adding other factors such as privacy concerns and unauthorized use. Although not specifically a cultural study, this work does highlight a potential use of mobile phones (i.e., emergency warning via phone) that may occur within or across culturally diverse areas.

Patterns of Usage

There are examples of patterns of usage affected by cultural dimensions. Sarker and Wells (2003) give an example of how the meaning of text messaging over mobile devices can differ depending on culture. For example, in Korea, which has a high power distance culture, texting to work supervisors (i.e., individuals higher in the "power" hierarchy) can be seen as offensive behavior. Users in Norway, a low power distance culture, did not see this texting as offensive, although the users thought it might be inappropriate for formal communications. Kayan, Fussell, and Setlock (2006) found that the use of instant messaging differed for Western individualistic, low-context cultures and Eastern collectivistic, high-context culture. Multi-party chat, audio-video chat and emoticons were more popular in the Eastern cultures than in the Western culture.

The preceding paragraphs describe only a few examples of the kinds of applications for which mobile devices are being used. There are certainly more potential application areas for mobile devices and the impact of culture needs to be an important consideration in developing and using such applications.

3.3. Mobile User Interface Design and Culture

Cultural dimensions have been used as ways to describe impacts on user interface design. For example, Baumgartner (2003) identified five dimensions ranked as most important in designing user interfaces: (1) context (Hall 1959; 1976); (2) technological development, referring to the rate of development as advanced to backward; (3) uncertainty avoidance (Hofstede, 1994); (4) time perception as described by Hall (1959; 1976) and (5) power distance (Hofstede, 1994).

Bossard (2008) also includes concrete guidelines for mobile phone user interfaces depending upon cultural background. Bossard used these guidelines to develop adaptive interfaces for various cultures as described by Hofstede's cultural dimensions. This approach took into account the small screen displays on mobile phones. The concrete guidelines are presented in table 3. Choi et al. (2005) conducted a study using interviews of users from Korea, Japan, and Finland where the users discussed their thoughts about video clips of mobile data services. The use cases chosen were downloading ring tones, downloading games, reserving movie tickets, and reading sports news. A total of 52 phone design features were extracted from interview data. Of those 52 features, 11 were found to be most important, being mentioned at least once by at least 90% of each of the three groups of respondents. They found that users from the same country showed similar likes and dislikes in design attributes of mobile data services. Results of the interviews were compared against the four cultural dimensions of uncertainty avoidance, individualism vs. collectivism, context, and time perception. There appeared to be relationships between the design attributes the users liked and characteristics (i.e., dimensions) of their cultures. The definitions of the eleven attributes and identification of which attributes were preferred or not preferred by the three nationalities are presented in table 4. The only one of the attributes that was mentioned and preferred by all 24 participants was "minimal steps or keystrokes." This suggests that this attribute can be generalized as important for all three nationalities, and might possibly be generalized across more, even most, cultures.

Table 4. Concrete guidelines for mobile phone user interfaces associated with cultural dimensions (adapted from Bossard, 2008, pages 14–15)

General Guidelines	Hofstede Classification[a]		Concrete Guidelines
Show breadcrumb trail	LTO	Low	Show breadcrumb trail. Generally, show position in the hierarchy.
Much information at first level, only a few levels	LTO PDI	High High	--Show a lot of menu items at the first level of navigation. --Use a shallow navigation hierarchy, or possibly no hierarchy.
Show only limited information at the first level, hierarchy can be deep.	LTO PDI	Low Low	--Show only a few menu items at the first level of the navigation. --Use a deep navigation hierarchy.
Let the system initiate communication.	PDI	High	If activated, help text is shown automatically.
User can initiate communication.	PDI	Low	If activated, help text is shown after the user presses a button.
Low image-to-text ratio	IDV	High	Use menus that can be shown without icons.

Table 4. (Continued)

General Guidelines	Hofstede Classification[a]		Concrete Guidelines
High image-to-text ratio	IDV	Low	Use menus that have icons as an inherent property. If a breadcrumb trail is used, show one with icons.
Contrast between interface areas	IDV	High	Use background color for contrasts. Possible interface areas could be the title bar, the breadcrumb trail, the area containing the menu items and the menu bar which contains the command labels. Menu items belonging together could be grouped by using the same background color.
Apply a high multimodality to the interface	IDV	Low	Use additional ouput mode possibilities, like sounds and vibration, to give the user cues about how deep inside the navigation he is.
Complex interface possible	UAV	Low	Use menus with a lot of menu items and add a lot of user interface components like icons and breadcrumb trail.
Simple user interface	UAV	High	Use menus with only few menu items; as few user interface components as possible.
Show the position of the user in the hierarchy	UAV	High	Use menus with breadcrumb trails. Maybe use a tree menu.
Use redundant cues	UAV	High	For example, use a different background color and bold font to signal the importance of a navigation item.
Use non-redundant cues to maximize information	UAV	Low	For example, use only bold font for the meaning "important". The background color can signal other meanings.
Show large amount of information within a screen, use the screen space efficiently	UAV	High	Use as many user interface components as possible. For example, use icons for navigation items to make their meaning clearer and use breadcrumb trails.

General Guidelines	Hofstede Classification[a]		Concrete Guidelines
Show secondary information	UAV	High	For example use help texts or icons to give more information about choices.
Show only limited choices	MAS UAV PDI	High High High	Try to limit the number of choices as much as possible. For example, a menu with maximum of 6 navigation items.

[a]This classification uses Hofstede's five cultural dimensions, with low and high values for the dimensions.
LTO = Long-term orientation.
PDI = Power distance.
MAS = Masculinity vs. femininity.
IDV = Individualism vs. collectivism UAV = Uncertainty avoidance.

Kim and Lee (2005) conducted a study looking at differences between Koreans and Americans in their ability to identify abstract vs. concrete icons and their preference for the icons. Americans recognized the abstract icons more than did the Koreans, while the Koreans recognized the concrete and semi-concrete icons more than the Americans. The authors conclude that there should be further studies to make clear the relation between culture and mobile phone interface design.

Lee et al. (2005) identified three important uses of cross-cultural usability data: (1) to determine global components and components that need to be localized; (2) to avoid cultural mistakes that might have an impact on use, such as miscommunication; and (3) to assess usability of cross-cultural user interfaces.

Using 28 participants (16 U.S. and 12 international cell phone users), a laboratory-based usability test was conducted to assess cell phone menu navigation system prototype usability.

Although the authors recognize that such a relatively small sample cannot represent entire national cultures, cross-culture information on navigation, metaphors, and appearance were gathered using a systematic methodology including recorded performance, verbal protocol analysis, and interview data. Prior to this study, there were not many systematic studies of mobile device use and the influence of culture.

Table 5. The eleven main attributes that partici pants felt were most important and their relationship with four cultural dimensions (Choi et al., 2005)

Attribute	Description	Uncertainty Avoidance		Individualism vs. Collectivism		Context		Time Perception (monochronic vs. polychromic)	
—	—	High	Low	Individ.	Collect.	High	Low	Mono	Poly
Minimal steps or keystrokes	Service requires minimal steps of keystrokes in search for desired contents.	All[a]	All	All	All	All	All	All	All
Iconic menu style	Menu items are represented by icons.					K, J[b] preferred	F not preferred	K,J,F preferred	
Secondary information about contents	Service provides additional information about contents, such as ranking points, movie ratings, or related pictures.	K, J preferred	F not preferred		K preferred				
Variety of contents	Service provides a wide assortment of contents to choose from. For example, for games, number of games, game genres; for movie ticket reservations, number of movie theaters; for ring tone downloads, number of ring tones, musical genre; for sports news, number of news items.	K, J preferred	F not preferred	F, J not preferred	K preferred				
Logical ordering of menu items	Service orders menu items or contents logically.	—	—	—	—	—	—	K,J,F preferred	

Attribute	Description	Uncertainty Avoidance		Individualism vs. Collectivism		Context		Time Perception (monochronic vs. polychromic)	
—	—	High	Low	Individ.	Collect.	High	Low	Mono	Poly
Clear menu labeling or hyperl i nk connotation	M enu labels tell users clearly what they can find behind them.	K, J preferred							
Efficient layout or space usage	Service uses the small screen space of the mobile phone efficiently.	K, J preferred							
Variety of font sizes	Services menus use a variety of font sizes for different menu types.				—	K, J preferred			
Variety of font colors	Service uses different colors for different menu items.				—	K, J preferred	F not preferred		
Large amount of information within a screen	Service displays a large amount of information on a given screen.	K, J preferred							
Various options for contents	Service provides various options for contents. For instance, for games, stage option, difficulty option; for move ticket reservations, seat selection; for ring tone downloads, chord selection; for sports news, choice of amount of information to display.	—	—	F, J preferred					

[a]All 24 participants said this was important bK = Korean, J = Japanese, F = Finnish.

Cha et al. (2005) conducted two studies to examine differences in mobile phone user interface preferences between people from the United Kingdom (Great Britain) and Korea. In the first, questionnaire-based study, they found that there were differences in preferences between the two nationalities. Korean users preferred Samsung while British users preferred Nokia phones. They used different features on the phones, with British users preferring talking and were not as interested in personalizing their phones with wallpaper or ringtones, while Korean users preferred texting and personalizing their phones. The second, user evaluation study showed that most Koreans users were frustrated with the simplicity of features while British users were frustrated with features that were too cluttered. Other differences in preferences were noted. However, both groups preferred fewer clicks, conducting easy tasks, and minimizing mistakes.

Frandsen-Thorlacius et al. (2009) discuss the concept of usability and whether or not it is constant across cultures. Their study had 412 users from China and Denmark complete questionnaires about how they understand and prioritize different aspects of usability. Based on the results, Chinese respondents identify visual appearance, satisfaction, and fun as being more important than the Danish participants, and the Danish place more importance on effectiveness and lack of frustration. Therefore, the authors conclude that cultural background has an influence on the perception of usability and usability cannot be assumed to be constant across cultures. The implications are that future assessments of usability across cultures need to think carefully about how usability is defined and assessed. An interesting point made is that previous usability research is predominantly Western in origin and therefore may contain "subtle but systematic cultural biases." (p. 41). Although this work is not specific to mobile device HCI usability, it clearly applies to mobile devices and suggests the influence of culture on definitions of usability.

In more recent literature, there is a shift from usability to user experience. This shift entails less focus on how the user interacts with the interface to more focus on the experience the user has, including not only practical functional requirements but also aspects to focus on user stimulation and identify needs (Walsh, Nurkka, Walsh, 2010). Kamppuri (2011) also has an extensive discussion about the change from usability to user experience in HCI research. Walsh et al. (2010) investigated whether user experience differed across cultures. To examine this question, the researchers gathered data using a remote online sentence completion survey, where users completed sentences describing various experiences. The authors concluded the users from different countries answered in different styles as well as content. The authors also were able to generate product improvement ideas from the survey results.

There are two views of design for a world-wide market—globalization (or internationalization) and localization. Globalization refers to the general process of worldwide economic, political, technological and social integration, and includes business interests such as making a product global (Jhangiani and Smith-Jackson, 2007). Localization refers to identifying local requirements, such as local cultural needs, and designing devices for local use. Eune and Lee (2009) define these views as "Internationalization is the process of designing software applications that it [sic] can be adapted to various languages and regions. Localization is the process of adapting internationalized product/service for a specific region or language..." (p. 2). Yan (2010), for example, advocates for understanding local cultures and designing different user experiences for people from different local cultures. The research is looking for strategies to develop multiple (i.e., localized) user interfaces using "cultural templates" or perhaps "special versioning tools." Interesting, Eune and Lee (2009) uses a word "glocalization" which the authors footnote as first appearing in the 1980s in articles by Japanese economists. The authors state, "Glocalization, a combined word globalization with localization [sic], is a term that was invented in order to emphasize that the globalization of a product is more likely to succeed when the product or service is adapted specifically to each locality or culture it is marketed in." Vyncke and Brengman (2010), after reviewing 27 relevant, high quality papers from the previous 12 years, found that culturally congruent websites had a positive impact on measures of "usefulness," "ease of use," positive attitudes," "positive intentions," and "overall effectiveness." While this paper did not address mobile devices, it can be postulated that such a positive effect would be found for culturally congruent websites on mobile devices as well.

As mentioned in the previous section, a challenge for using mobile phones in the developing world is the challenge of creating interfaces for low-literacy or novice users. Medhi et al. (2011) has identified alternative input and output methods for mobile interfaces (see figure 9). The authors have conducted two studies, both in India, that compared various methods of input and output. For the study that evaluated text versus voice and graphics, with non- or semi-literate participants, the text interface was unusable by individuals if they were non-literate. Spoken dialogue and graphical output both had advantages and disadvantages. Non-literate users need non-text interfaces, and spoken input with graphical output may combine benefits of each approach. A second study had literate health workers entering health data by either text entry or speaking with a live operator. Live operators decreased errors and the recommendation is to use live operator, if economically feasible. While low literacy is not unique to any culture, it is a challenge that needs to be considered in cross-cultural environments.

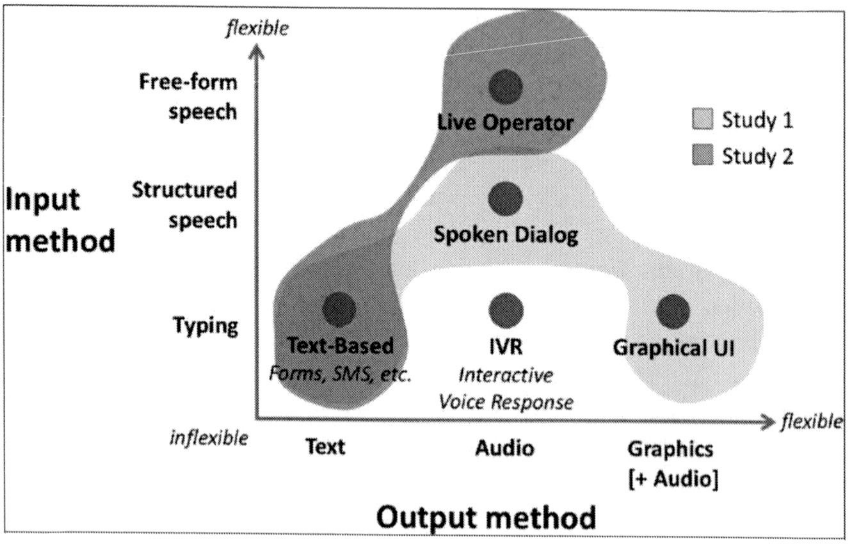

Figure 9. Design space of mobile user interfaces. Shaded regions indicate the comparative studies presented in the Medhi et al. (2011) article. (Medhi, Patnaik, Brunskill, Gautama, Thies and Toyama, 2011).

Jhangiani and Smith-Jackson (2007) looked at groups in the U.S. and India, without any disability or with a visual disability (i.e., legally blind.) They were interested in comparing two kinds of culture—national culture and disability culture. They found that there were differences in usability ratings, some based on the disability group, some on the national groups and for some mobile phone functions for the disability specific nationality groups. This finding emphasizes the need to consider the influence of culture on usability. The authors also propose a model for cell phone design adapted from Jordan's pleasurability framework (Jordan, 2000). The model, shown in figure 10, relates the four pleasures from Jordan to mobile cell phone design requirements. It is interesting to note that the elements in the model are all elements that may influence use differently in different cultural contexts; for this particular paper, the focus is on disability and nationality cultures but could also be any other cultural context.

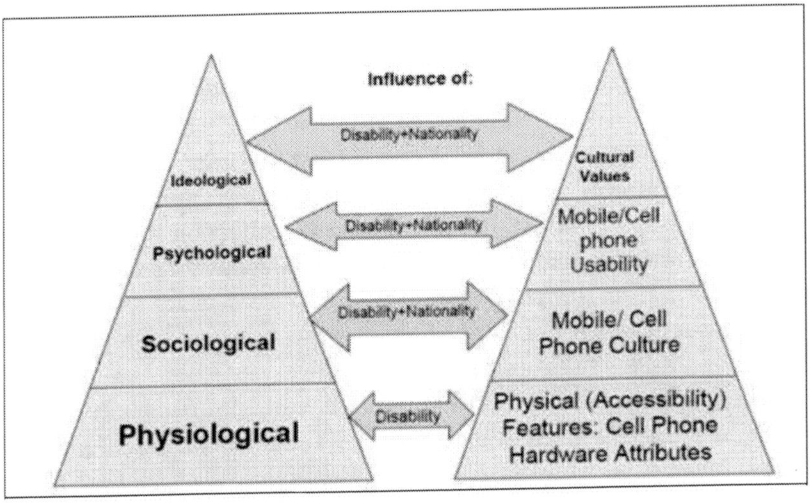

Figure 10. Pyramid model for mobile cell phone design (adopted from Jordan (2000) (from Jhangiani and Smith-Jackson, 2007).

4. DISCUSSION

As shown by the research discussed in this report, there is a great amount of interest in the use of mobile technology, specifically mobile phones, and the influence and impact of culture on the adoption, use and design of the technology.

4.1. Observations

An observation from this literature review is that the applicable research literature has been obtained from many sources and disciplines. A number of articles are from the "traditional" HCI literature. However, there are a number of articles from information sciences and systems literature. Given the interest in mobile commerce and business opportunities, there may be additional relevant literature in more business- and marketing-oriented sources. Another observation is that several recent dissertations/thesis address the topic of interest (e.g., Kamppuri, 2011; Matsui, 2009; Bossard, 2008; van Biljon, 2006) which suggest the timeliness and emerging nature of this topic and how it is not yet well understood.

A primary purpose of this literature review was to identify factors that are important in understanding the adoption decision and the use of mobile devices and the impact of cultural elements on the decision to adopt and use. If we can identify factors important to the influence of culture on mobile device adoption and usage, then these same factors may be applicable, more generally, to influences on communications and decision making—topics that are more generally of interest. Factors important to adoption and use may also provide some insights that would be valuable in supporting the creation of guidelines for system design, specifically for mobile devices. As seen in this report, there is research literature addressing the design of mobile interfaces for various cultures. Such guidelines may prove useful for U.S. Army and multinational military as interest increases in the use of smartphones and other lightweight, minimally sized display interfaces for military applications.

This review has been successful in identifying a number of relevant factors that are related to the adoption and use of mobile devices. There are several models of technology acceptance and adoption that are continually being revised and refined for particular focused research questions. Of particular interest are the models of mobile technology adoption and use that incorporate cultural factors (e.g., van Biljon and Kotze, 2008; Min, Li and Ji, 2009). It is difficult to say if the models are complete; in fact, they are almost surely not complete. But as a first approximation, the models may prove to be very useful in identifying factors relevant to cultural impacts on other areas of human performance and interaction with technology, such as decision making and communication. Potential factors of influence on decision making and communications can be drawn from these models, as both are present implicitly in the areas of making decisions to adopt and to use (e.g., communicate with) mobile technology.

We can identify some potentially relevant factors from this examination of the impact of culture on adoption, design and use of mobile devices as we explore the impact of culture on, more generally, decision making and communications. Successful decision making and communications are instrumental for effective military operations. An initial list of important factors that may influence decision making and communications, drawn from this literature review, include: (1) demographics, (age; gender, income); (2) personality factors (e.g., innovativeness, uncertainty avoidance; apprehensiveness); (3) attitudes (e.g., expectations; perceived usefulness); and, (4) social influence. These are all factors that have been identified by one or more of the models discussed in this report. They are all factors that affect adoption decisions and use of mobile devices for communications and other tasks. They also are areas that may be impacted by culture—in some of the models of adoption and use, culture is explicitly identified while in other models, it is not addressed at all.

4.2. Relevance to Army

The topic of mobile technology is very relevant to the Army. There is a move to providing mobile phones and software to Soldiers (Lane, 2010; Ackerman, April 22, 27, 2011.) Information we can learn about appropriate and effective design of mobile services and applications will be useful to the Army. There is considerable information available on design guidance for culturally appropriate user interfaces and experiences. This information can be used in design of software for U.S. Soldiers and for other national forces who may be our coalition partners.

The U.S. Army has interest in mobile devices as a means for communications and as a way to disseminate applications (i.e., apps.) A successful app can be thought of as "an elegant, easy-to use application that solves a real problem or provides meaningful entertainment." (Barnard et al 2009; p. 3). For the Army, we can think of providing software applications that solve real problems. (We will not consider the entertainment apps here.) The Army has adopted the app approach through competitions (e.g., U.S. Army Times, 2010; Lopez, 2010) and its own app store (Ackerman, April 2011; U.S. Army Architecture Community, 2010). It is possible to conceive of Army apps yet to be designed that can aid in mission accomplishment, for example, apps that address issues with local populations or humanitarian missions.

A recent article in the *Army AL&T* (*Acquisition, Logistics and Technology*) magazine (Daniels and Stiner, 2011) describes a test where smartphones were plugged into tactical radios, allowing Soldiers to communicate using voice, data, or images. The authors suggested that the use of the smartphones and associated technology allowed the Soldiers "to be faster, quieter, and therefore less detectable—all of which can save lives" (p. 31) because the Soldiers did not need to communicate by yelling across distances. These ruggedized Android-based phones ran two apps: the Joint Battle Command-Platform (JBC-P Handheld) and the Tactical Ground Reporting (TIGR). One benefit of this technology cited was the small amount of training needed because of Soldier familiarity with cell phones. Another interesting example of possible benefits is that Soldiers could take a photo of a wound and send it to a doctor for assistance in getting precisely appropriate treatment. Within the article, it is reported that smartphones will continue to be pursued for use by Army program managers (Daniels and Stiner, 2011).

We know that mobile devices, and smartphones in particular, are ubiquitous in our society. Mobile smartphone technology is used within the Army. These technologies are not only used within the U.S., but are used globally. What is not known at this time is how culture and mobile technology (including software design) can be exploited to serve Army missions abroad. Can mobile apps be pushed to the local population? Can information be gathered from the local population via mobile technology? There may be many ways to use this technology in service to Army goals that have not yet been imagined or fully developed. Can this technology be used to further positive interaction and success with other national forces with whom we interact and partner?

Several specific research topics seemed of practical interest for the Army for the future. The first is the development of multiple, culturally appropriate versions of user interfaces and services through cultural templates or special versioning tools, as explored by Yan (2010). If the Army has a need for applications specific to cultures, this may be a useful approach. Similarly, the idea of an adaptive interface, either directly changing the interface, as suggested by Bossard (2008), or through an adaptive dynamic agent for international collaboration which adapts its social interaction to the user's cultural background, as Evers et al. (2010) has done, could each be used to develop and adapt applications for specific cultures.

Another intriguing concept is the use of mobile devices as a distributed sensor and communications network, where information from humans can be received and then integrated with location-based information to provide a (potentially) robust description of a local environment. Mobile phones might also be used as a distributed, and potentially asynchronous, communications tool, used to interrupt, ask questions, and send instructions, as Evers et al. (2010) demonstrated. Other culturally appropriate and effective communication uses that would support Army missions could be imagined; for example, mobile tools to support communications between U.S. and national forces.

In summary, this literature review only scratches the surface in identifying and reviewing studies related to how culture impacts the adoption, use, and design of mobile technology. We do not claim that this review is exhaustive. However, we hope the information and observations here may be valuable in furthering Army missions.

REFERENCES

Ackerman, S. First Look: Inside the Army's App Store for War, April 27, 2011. http://www.wired. com/dangerroom/2011/04/armys-app-store-for-war (accessed 20 June 2011).

Ackerman, S. U.S. Army picks Android to power its first smartphone. Arstechnica.com.. April 22, 2011. http://arstechnica.com/gadgets/news /2011/04/us-army-picks-android-to-power-its- first-smartphone.ars?utm_ source=feedburner&utm_medium=feed&utm_campaign=F (accessed 9 June 2011).

Aloudat, A.; Michael, K. Toward the Regulation of Ubiquitous Mobile Government: A Case Study on Location-Based Emergency Services in Australia. *Electronic Commerce Research,* 2011, *11* (1), 31–74.

Aryana, B.; Boks, C. Cultural Customization of Mobile Communication Devices' Components. *International Design Conference (Design 2010),* Dubrovnik, Croatia, 17–20 May 2010, pp 137–146.

Aryana, B.; Oritsland, T. A. Culture and mobile HCI: A review. *NordDesign 2010*, 25–27 August 2010, Goteborg, Sweden. http://s3images. coroflot. com/user_files/individual_files/92608_zuA873FXC6UyiMS3r8unZKUZt. pdf (accessed 25 May 2011).

Bagchi, K.; Kirs, P. The Impact of Schwartz's Cultural Value Types on ICT use: A Multi-National Individual-Level Analysis. *International Conference on Information Systems (ICIS) 2009 Proceedings* Paper 205, http://aisel.aisnet.org/icis2009/205.

Barnard, D.; Bondo, J.; Burcaw, D.; Kaneda, D.; Kemper, C.; Novikoff, T.; Parrish, C.; Ellis, B.; Peters, K.; Siebert, J.; Wilson, E. *iPhone User Interface Design Projects*. NY: Apress, 2009.

Baumgartner, V-J. *A Practical Set of Cultural Dimensions For Global User-Interface Analysis and Design*, June 2003.. Diploma Thesis, Fachhochschule Joanneum, Graz, Austria. http://www.mavas.at/val /education05_thesis00.asp (accessed 12 June 2011).

Blom, J.; Chipchase, J.; Lehikoinen, J. Contextual and Cultural Challenges for User Mobility Research. *Communications of the ACM,* 2005, *48* (7), 37–41.

Bossard, A. *Ontology-Based Cultural Personalization in Mobile Applications*. Diploma Thesis, University of Zurich, pp 111, 2008.

Cha, H.; Oshlyansky, L.; Cairns, P. Mobile Phone Preferences and Values: The U.K. vs. Korea. In *International Workshop on Internationalisation of*

Products and Systems (IWIPS). http://www-users.cs.york.ac.uk /~pcairns/papers/IWIPS_2005.pdf (accessed 24 August 2011).

Choi, B.; Lee, I.; Kim, J.; Jeon, Y. A Qualitative Cross-National Study of Cultural Influences on Mobile Data Service Design. *Proceedings of the SIGCHI Conference on Human Factors in Computing Systems,* CHI '05, ACM, 2005;pp 661–670.

Constantiou, I.; Papazafeiropoulou, A.; Vendelo, M. Does Culture Affect the Adoption of Advanced Mobile Services? A Comparative Study of Young Adults' Perceptions in Denmark and the UK. *The DATA BASE for Advances in Information Systems* 2009 *40* (4), 132–147.

Davis, F.; Bagozzi, R.; Warshaw, P. User Acceptance of Computer Technology: A Comparison of Two Theoretical Models. *Management Science* 1989 *35* (8), 982–1003.

Daniels, M.; Stiner, M. Information on the Fly. *Army AL&T Magazine*, July-September 2011, 30–33.

Donner, J. Research Approaches to Mobile Use in the Developing World: A Review of the Literature. *The Information Society* 2008, *24* (3).

Eune, J.; Lee, K. P. Analysis on Intercultural Differences through User Experiences of Mobile Phone for Glocalization [sic]. *Proceedings of the International Association of Societies of Design* (IASDR 2009).(Coex, Seoul, Korea, 18–22 October 2009). http://www.iasdr2009. org/ap/Papers/Orally%20Presented%20Papers/Culture/Analysis%20on%2 0Intercultural%20
Differences%20through%20User%20Experiences%20of%20Mobile%20P hone%20for%20gl ocalization.pdf (accessed 25 May 2011).

Evers, V.; Maldonado, H.; Brodecki, T.; Hinds, P. Relational vs. Group Self-Construal: Untangling the Role of National Culture in HRI. In T. Fong and K. Dautenhahn Eds; HRI pp 255–262 New York: NY, 2008; ACM.

Evers, V.; Winterboer, A.; Pavlin, G.; Groen, F. *ICIC10*, Copenhagen, Denmark, 19–20 August pp 183–186, 2010 ACM.

Frandsen-Thorlacius, O.; Hornbaek, K.; Hertzum, M.; Clemmensen, T. Non-Universal Usability? A Survey of How Usability Is Understood by Chinese and Danish Users, *Proceedings of CHI 2009*, pp 41–50, 2009. ACM.

Guerrero, L.; Ochoa, S.; Pino, J. Selecting Computing Devices to Support Mobile Collaboration. Group *Decision and Negotiation* 2006, *15*, 243–271.

Hall, E. *The Silent Language*. NY: Doubleday, 1959.

Hall, E.T. *Beyond Culture.* Garden City, NY: Anchot Press, 1976.

Herskovic, V.; Ochoa, S.; Pino, J.; Neyem, A. The Iceberg Effect: Behind the User Interface of Mobile Collaborative Systems. *Journal of Universal Computer Science,* 2011, *17* (2), 183–202.

Hofstede, G. *Culture's Consequences: International Differences in Work-Relate Values*. Beverly Hills, CA: 1980. Sage.

Hofstede, G. The Business of International Buisness is Cultur5228672. e. *International Business Review* 1994, *3* (1), 1–14.

Hofstede, G. The Business of International Business Is Culture. In Cross Cultural Management, T. Jackson Eds; London: Buterworth-Heinemann, 1995; pp 150–154.

Hofstede, G.; McCrae, R. Personality and Culture Revisited: Linking Traits and Dimensions Of Culture, *Cross-Cultural Research* 2004, *38* (1), 52–88.

Jhangiani, I.; Smith-Jackson, T. Comparison of Mobile Phone User Interface Design Preferences: Perspectives From Nationality and Disability Culture, *Proceedings of the 4th International Conference on Mobile Technology, Applications and Systems* (Mobility'07, 10–12 September, Singapore, pp 512–519, 2007 ACM.

Jordan, P. W. *Designing Pleasurable Products*. London: Taylor and Francis, 2000.

Kamppuri, M. *Theoretical and Methodological Challenges of Cross-Cultural Interaction Design.* (Doctoral dissertation, University of Eastern Finland, 2011). http://epublications.uef.fi/pub/urn_isbn_978-952-61-0407-2/urn_isbn_978-952-61-0407- 2.pdf (accessed 25 May 25 2011).

Kayan, S.; Fussell, S.; Setlock, L. Cultural Differences in the Use of Instant Messaging in Asia and North America. *CSCW'06 Proceedings of the 2006 20th Anniversary Conference on Computer Supported Cooperative Work*, pp 525–528, NY: 2006. ACM.

Kim, J. H.; Lee, K. P. Cultural Difference and Mobile Phone Interface Design: Icon Recognition According to Level of Abstraction. *Proceedings of MobileHCI'05*, Salzburg, Austria, 19–22 September 2005 pp 307–310; ACM.

Kluckhohn, F.; Strodtbeck, F. *Variations in Value Orientations*. Oxford, England: Row, Peterson, 1961.

Koo, C.; Wati, Y. Toward an Understanding of the Mediating Role of "Trust" in Mobile Banking Service: An Empirical Test of Indonesia Case. *Journal of Universal Computer Science* 2010, *16* (13), 1801–1824. http://www.jucs.org/jucs_16_13/toward_

an_understanding_of/jucs_16_13_1801_1824_koo.pdf (accessed 25 May 25 2011).

Kwon, H. S.; Chidambaram, L. A Test of The Technology Acceptance Model, the Case of the Cellular Telephone Adoption. *IEEE Proceedings of the 33*rd *Hawaii International Conference on System Sciences*, 2000; pp 1–10.

Lane, S. U.S. Army wants to equip all soldiers with an Apple iPhone or Android phone. *AppleInsider*. http://www.appleinsider.com /articles/10/12/14/us_ army_wants_to_give_soldiers_a_choice_of_apple_ iphone_or_android_phone.html (accessed 9 June 2010).

Lee, Y. S.; Ryu, Y. S.; Smith-Jackson, T. L.; Shin, D. J.; Nussbaum, M. A.; Tomioka, K. Usability Testing With Cultural Groups in Developing a Cell Phone Navigation System." *Proceedings of HCI International 2005 – 11th International Conference on Human-Computer Interaction*, Las Vegas, NV, 2005.

Lopez, C. T. Apps for Army to Shape Future Software Acquisition. http://www.army.mil/article /43293/apps-for-army-to-shape-future-software-acquisition (accessed 20 June 2011).

Lu, J.; Yao, J. E.; Yu, C. S. Personal Innovativeness, Social Influences and Adoption of Wireless Internet Services via Mobile Technology. *Journal of Strategic Information Systems,* 2005, 14 (3), 245–268.

Martin, B.; Abbott, E. *Development calling: The use of mobile phones in agriculture development in Uganda*. Presented at the International Federation for Information Processing workshop, "Information and Communications Technologies (ICT) and Development Research Voices from Africa," IFIP9, 22–23 March 2010, at Makerere University, Kampala, Uganda. http://mak.ac.ug/documents/IFIP/Devleopment Calling.pdf (accessed 20 June 2011).

Massey, A. P.; Khatri, V.; Ramesh, V. *From the Web to the Wireless Web: Technology Readiness and Usability. Paper presented at the 38th Annual Hawaii International Conference on System Sciences*, Big Island, HA, 2005.

Mathew, J.; Sarker, S.; Varshney, U. M-commerce services: Promises and challenges. *Communications of the Association for Information Systems,* 2004, *14* (26) (accessed 16 June 2011).

Matsui, K. *Influence of Cultural Difference on Mobile User Experience. A Case Study of the Nokia Phone in Japanese Mobile Culture. (*Master's thesis, University of Jyvaskyla, Finland, 2009.) https://jyx.jyu.fi/dspace /bitstream/handle/123456789/22463/URN_NBN_fi_jyu200911184415. pdf?sequence=1 (accessed 25 May 2011).

Medhi, I.; Patnaik, S.; Brunskill, E.; Gautama, S.N. N.; Thies, W.; Toyama, K. Designing Mobile Interfaces for Novice and Low-Literacy Users. *ACM Transactions on Computer-Human Interaction,* 2011, *18*, (1), 2.

Min, Q.; Ji, S.; Qu, G. Mobile Commerce User Acceptance Study in China: A Revised UTAUT Model. *Tsinghua Science and Technology* 2008, *13* (3), 257–264. http://qhxb.lib. tsinghua.edu.cn/myweb/english//2008/2008e3 /257-264.pdf (accessed 25 May 2011).

Min, Q.; Li, Y.; Ji, S. The Effects of Individual-Level Culture on Mobile Commerce Adoption: An Empirical Study. *IEEE 2009 Eighth International Conference on Mobile Business*, 2009; pp 305–312.

mobiThinking. Global Mobile Statistics 2011: All Quality Mobile Marketing Research, Mobile Web Stats, Subscribers, Ad Revenue Usage, Trends.... http://mobithinking.com/mobilemarketing-tools/latest-mobile-stats (accessed 20 June 2011).

Parsons, T.; Shils, E. *Toward a General Theory of Action*. Cambridge, MA: Harvard University Press, 1951.

Rosoff, M. Apple leaked how many iPads it sold in Q2 and it's a good number. Business Insider website, http://www.businessinsider.com/new-expectation-for-ipad-sales-8-millionthis-quarter-2011-6 (accessed 29 August 2011).

Sarker, S.; Wells, J. Understanding Mobile Handheld Device Use and Adoption. *Communications of the ACM,* 2003, *46* (12), 35–40.

Schwartz, S. H. Universals in the Content and Structure of Values: Theoretical Advances and Empirical Tests in 20 Countries. *Advances in Experimental Social Psychology*, 1992, 25, 1–65.

Schwartz, S. H.; Ros, M. Value in the West: A Theoretical And Empirical Challenge to the Individualism-Collectivism Cultural Dimension. *World Psychology* 1995, *1,* 99–122.

Trompenaars, F. *Riding the Waves of Culture*. London: Nicholas Brealy Publishing, 1993.

U.S. Army Architecture Community Introducing the U.S. Army Marketplace! http:// architecture.army.mil/technical-view/applications/applications.html (accessed 20 June 2011).

U.S. Army Times, 3 March 2010, Army launches app development contest. http://www.armytimes.com/news/2010/03/army_a4a_030310w/ (accessed 20 June 2011).

Van Biljon, J. A. *A Model for Representing the Motivational and Cultural Factors That Influence Mobile Phone Usage Variety*. (Doctoral dissertation, University of South Africa, 2006). http://uir.unisa.ac.za

/bitstream/handle/10500/2149/thesis.pdf?sequence=1 (accessed 25 May 2011).

Van Biljon, J.; Kotze, P. Modelling the Factors That Influence Mobile Phone Adoption. *Proceedings of the 2007 Annual Research Conference of The South African Institute of Computer Scientists and Information Technologists on IT Research in Developing Countries, SAICSIT'07*, NY, 2007a, pp 152–161; ACM.

Van Biljon, J.; Kotze, P. Profiling mLearning Students According to Cultural Dimensions: Is that possible? *Proceedings of the 37*th *Southern Africa Computer Lecturers' Association Conference,* 1–3 July 2007, Victoria Falls, Zimbabwe, pp 1–14, http://uir.unisa. ac.za/handle/10500/3137? show=full (accessed 14 June 2011).

Van Biljon, J.; Kotze, P. Cultural Factors in a Mobile Phone Adoption and Usage Model. *Journal of Universal Computer Science* 2008, *14* (16), 2650–2679.
http://www.jucs.org/jucs_14_16/cultural_factors_in_a/jucs_14_16_2650_2679_biljon.pdf (accessed 25 May 2011).

Venkatesh, V.; Morris, M.; Davis, G.; Davis, F. User Acceptance of Information Technology: Toward a Unified View, *MIS Quarterly* 2003, *27* (3), 425-478.

Walsh, T.; Nurkka, P.; Walsh, R. Cultural Differences in Smartphone User Experience Evaluation. *Proceedings of the* 9th *International Conference on Mobile and Ubiquitous Multimedia,* MUM '10, no. 24, 2010 ACM.

Wang, L.; Rau, P.; Evers, V.; Robinson B.; Hinds, P. Responsiveness to Robots: Effects of ingroup orientation & communication style on HRI in China. *Proceedings of Human Root Interaction Conference HRI'09,* San Diego, CA, 11–13 March 2009.

Wu, J. H.; Wang, S. C. What Drives Mobile Commerce? An Empirical Evaluation of the Revised Technology Acceptance Model. *Information and Management,* 2005, 42: 719–729.

Yan, Q. Cultural Versioning of Mobile User Experience. *Proceedings of the CHI 2010.* Atlanta, GA, 10–15 April 2010, pp 2943–2946; NY: ACM.

Yang, K. C. C. Exploring Factors Affecting the Adoption of Mobile Commerce in Singapore. *Telematics and Informatics* 2005, *22*, 257–277.

Vyncke, F.; Brengman, M. Are Culturally Congruent Websites More Effective? An Overview of a Decade of Empirical Evidence. *Journal of Electronic Commerce Research* 2010, *11* (1), 14–29.

List of Symbols, Abbreviations, and Acronyms

CSCW	computer-supported cooperative work
F	Finland
HCI	human-computer interaction
IDV	Individualism vs. collectivism
J	Japan
JBC-P	Joint Battle Command-Platform
K	Korea
LTO	Long-term orientation
MAS	Masculinity vs. femininity
mLearning	Mobile Learning
MOPTAM	Mobile Phone Technology Adoption Model
PC	personal computer
PDI	Power distance
SI	social influence
TAM	Technology Acceptance Model
TIGR	Tactical Ground Reporting
UA	uncertainty avoidance
UAV	Uncertainty avoidance
UTAUT	Unified Theory of Acceptance and Use of Technology

In: America's Embrace of the Mobile Internet ISBN: 978-1-63463-585-1
Editor: Jennifer Haggard © 2015 Nova Science Publishers, Inc.

Chapter 3

TEXT AND MULTIMEDIA MESSAGING: ISSUES FOR CONGRESS*

Patricia Moloney Figliola and Gina Stevens

SUMMARY

The first text messages were sent during 1992 and 1993, although commercially, text messaging was not widely offered or used until 2000. Even then, messages could only be sent between users subscribed to the same wireless carrier; for example, Sprint customers could only exchange messages with other Sprint customers. In November 2001, however, wireless service providers began to connect their networks for text messaging, allowing subscribers on different networks to exchange text messages. Since then, the number of text messages in the United States has grown to over 48 billion messages every month. Additionally, text messages are no longer only sent as "point-to-point" communications between two mobile device users. More specifically, messages are also commonly sent from web-based applications within a web browser (e.g., from an Internet e-mail address) and from instant messaging clients like AIM or MSN.

For congressional policymakers, two major categories of issues have arisen: (1) "same problem, different platform" and (2) issues stemming from the difficulty in applying existing technical definitions to a new

* This is an edited, reformatted and augmented version of a Congressional Research Service publication RL34632, prepared for Members and Committees of Congress, dated January 12, 2012.

service, such as whether a text message is sent "phone-to-phone" or using the phone's associated email address. There are numerous examples of each. An example of the first category would be consumer fraud and children's accessing inappropriate content, which have existed previously in the "wired world," but have now found their way to the "wireless world." An example of the second category would be that spam sent between two phones or from one phone to many phones does not fall under the definition of spam in the CAN-SPAM Act of 2003 (Controlling the Assault of Non-Solicited Pornography and Marketing Act, P.L. 108-187); however, if that same message were to be sent from a phone or computer using the phone's associated e-mail address, it would.

The increasing use of text and multimedia messaging has raised several policy issues: distracted driving, SMS spam, the inability of consumers to disable text messaging, text messaging price fixing, carrier blocking of common short code messages, deceptive and misleading common short code programs, protecting children from inappropriate content on wireless devices, "sexting," mobile cyberbullying, privacy of text messages, and using SMS to support law enforcement and emergency response.

INTRODUCTION

The first text messages were sent during 1992 and 1993, although commercially, text messaging was not widely offered or used until 2000. Even then, messages could only be sent between users subscribed to the same wireless carrier; for example, Sprint customers could only exchange messages with other Sprint customers. In November 2001, however, wireless service providers began to connect their networks for text messaging, allowing subscribers on different networks to exchange text messages. Since then, the number of text messages in the United States has grown to over 48 billion messages every month. Additionally, text messages are no longer only sent as "point-to-point" communications between two mobile device users. For example, messages are also commonly sent from web-based applications within a web browser and from instant messaging clients like AIM, MSN, or Google Chat.

DEFINITIONS

Short Message Service

Short Message Service (SMS) is a method of communication that sends text between cell phones, or from a computer or handheld device to a cell phone. The "short" part refers to the maximum size of the text messages: 160 characters.[1] The term "SMS" is generally used interchangeably with the term "text message."

Even when not being used for a voice call, a mobile phone is constantly sending and receiving information. It is communicating to its cell phone tower over a control channel. The reason for this communication is so that the cell phone system knows which cell a phone is in, and so that the phone can change cells as the user moves around. Every so often, a phone and a tower will exchange a packet of data that lets both "know" that everything is working properly.

The control channel also provides the pathway for SMS messages. When someone sends an SMS message, the message flows through the SMS Center (SMSC), then to the cell tower, and the tower then sends the message to the recipient's phone as a packet of data on the control channel. *Figure 1* illustrates how an SMS message is processed.

Enhanced and Multimedia Message Service

While SMS only allows plain text to be sent, two alternative messaging services allow for more elaborate types of messages. With Enhanced Messaging Service (EMS), formatted text, sound effects, small pictures, and icons can be sent. MMS (Multimedia Messaging Service) allows animations, audio, and video files in addition to text to be sent.

E-mail-to-SMS Messaging

As noted above, SMS messages may be sent between a computer and a mobile phone. However, these messages are sent using the e-mail address associated with the mobile device, such as 2025551212@carrier.com. For that reason, these messages are classified as e-mail and therefore are subject to different and more stringent regulation (see "SMS Spam").

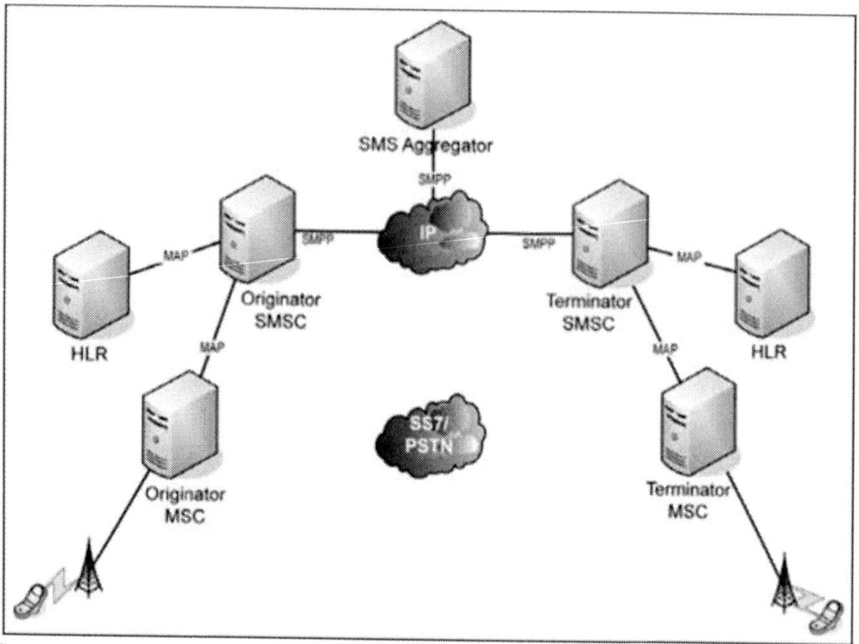

Source: Used with permission from Motorola. Definitions: The "Internet Protocol (IP) cloud" represents an Internet Protocol network used to carry data traffic; HLR = Home Location Register (the central database that contains details of each mobile phone subscriber); MAP = Mobile Application Part signaling protocol; MSC = Mobile Switching Center; the "Public Switched Telephone Network (PSTN) cloud" is included to demonstrate that SMS messages are not carried over it; SMS Aggregator = an intermediary between mobile service providers providing SMS service; SMSC = SMS Center; SMPP = Short Message Peer-to-Peer Protocol.

Figure 1. Path of Intercarrier SMS Messages.

Common Short Codes (CSCs)

Introduced in the U.S. market in October 2003, Common Short Codes (CSCs) are short numeric codes of five or six digits, compatible across carriers, to which text messages can be sent from a mobile phone. Wireless subscribers send text messages to short codes to access a wide variety of mobile content, for example, to vote for contestants on American Idol. Many entities use CSCs to communicate with interested parties: television stations; individual television shows; radio stations; instant messaging services; political, advocacy, and other organizations; magazines; and sports teams—

among others. Users send a message to the CSC to subscribe to alerts or other messages. Sometimes these messages are delivered for free by the originator, sometimes there is a fee. *Figure 2* illustrates how a CSC message is processed.

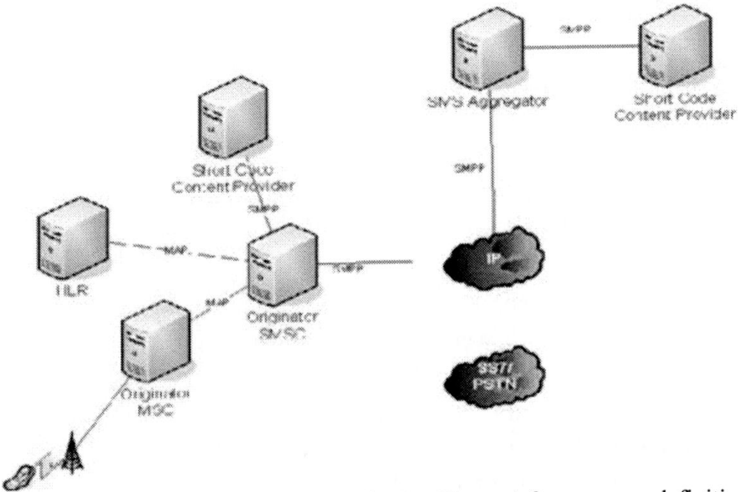

Source: Used with permission from Motorola. See Figure 1 for acronym definitions.

Figure 2. Path of Common Short Code Messages.

"Vanity" CSCs are also available (for a higher price)—these CSCs use letters on a mobile device keypad to spell out words that are easy to remember and are chosen to reflect the service the short code is being used to access.[2] Furthermore, although CSCs can be "compatible" across all carriers, some CSCs are established as business partnerships between a specific carrier and another entity. For example, American Idol has an exclusive partnership with AT&T Wireless.[3]

ISSUES FOR CONGRESS

For congressional policymakers, the major issues that have arisen stem from what could be called "same problem, different platform." For example, issues such as consumer fraud and children's accessing inappropriate content, which have existed previously in the "wired world," have now found their way to the "wireless world."

Other issues stem from the difficulty in applying technical definitions to a given service, such as whether a text message is sent "phone-to-phone" or using the phone's associated e-mail address. For example, spam sent between two phones or from one phone to many phones does not fall under the legal definition of spam; but if that same message is sent from a phone or computer using the phone's associated e-mail address, it does.

Distracted Driving Caused By Texting

According to the U.S. Department of Transportation (DOT), approximately 16% of fatalities in distraction-related crashes were caused in at least some part by mobile devices. Further,

- In 2009, 5,474 people were killed in U.S. roadways and an estimated additional 448,000 were injured in motor vehicle crashes that were reported to have involved distracted driving.
- The age group with the greatest proportion of distracted drivers was the under-20 age group—16% of all drivers younger than 20 involved in fatal crashes were reported to have been distracted while driving.
- Drivers who use hand-held devices are four times as likely to get into crashes serious enough to injure themselves.
- Using a cell phone while driving, whether it is hand-held or hands-free, delays a driver's reactions as much as having a blood alcohol concentration at the legal limit of 0.08 percent.[4]

While reading and composing text messages while driving is only one of numerous factors that can lead to distracted driving, such activity is a growing concern among safety and regulatory groups. In response to this concern, there have been various actions taken at the federal and state levels.

Legislation—112th Congress

A number of bills have been introduced in the 112th Congress on the issue of "distracted driving."

Motorcoach Enhanced Safety Act of 2011, H.R. 873 Motorcoach Enhanced Safety Act of 2011, S. 453 Commercial Motor Vehicle Safety Enhancement Act of 2011, S. 1950

The Motorcoach Enhanced Safety Act was introduced in the Senate by Senator Sherrod Brown and in the House by Representative John Lewis on

March 2, 2011. With respect to text messaging, the Motorcoach Enhanced Safety Act would require the Secretary of Transportation to

- prescribe regulations on the use of electronic or wireless devices (including cell phones and other distracting devices) by motorcoach operators; and
- prohibit their use in cases where they interfere with the driver's safe operation of a motorcoach, but not when necessary for driver or public safety in emergency situations.

S. 453 was referred to the Senate Committee on Commerce, Science, and Transportation on December 14, 2011, and ordered to be reported with an amendment in the nature of a substitute favorably. The text of S. 453 was included in Commercial Motor Vehicle Safety Enhancement Act of 2011, S. 1950. That bill was introduced by Senator Frank Lautenberg on December 7, 2011;[5] the bill was ordered to be reported with an amendment in the nature of a substitute favorably.

Mariah's Act, S. 1449

Senator Mark Pryor introduced Mariah's Act on July 29, 2011. The bill would create grants to states that enact and enforce a law that

- prohibits drivers from texting while driving;
- prohibits drivers age 18 or younger from using a cell phone while driving, making violation of the law a primary offense; and
- establishes certain minimum fines and increased civil and criminal penalties. The bill would also create a distracted driving grant program at the National Highway Traffic Safety Administration (NHTSA).

The bill was referred to the Committee on Commerce, Science, and Transportation. On December 14, 2011, the committee ordered the bill to be reported with an amendment in the nature of a substitute favorably.

Students Taking Action for Road Safety Act (STARS Act) of 2011, S. 1422

Senator Amy Klobuchar introduced this bill on July 27, 2011. The bill focuses on teen traffic safety in general, including providing grants to educate teens on issues related to safe driving. Specifically, the bill:

- Directs the Secretary of Transportation to establish a teen traffic safety grant program to award formula grants to states to implement statewide programs to improve the traffic safety of teen drivers.
- Authorizes a state to use grant funds to implement such statewide program to improve the traffic safety of teen drivers, including activities to support peer-topeer education and prevention strategies in schools and communities to increase safety belt use and reduce speeding, impaired and distracted driving, underage drinking, and other destructive teen driver decisions that lead to injuries and fatalities.
- Authorizes the Secretary to contract with a national, nonprofit organization to establish a technical assistance center to provide training and technical assistance to state and local officials, student leaders, school advisors, and other entities associated with the grant program. Authorizes the center to operate a national teen traffic safety clearinghouse.
- Directs the Secretary to establish the National Teen Driver Advisory Council to study and develop an education and prevention strategy to reduce teen driver injuries and fatalities.

The bill was referred to the Committee on Commerce, Science, and Transportation. No further action has been taken.

Safe Drivers Act of 2011, H.R. 2333

Representative Carolyn McCarthy introduced this bill on June 23, 2011. The act would:

- Direct the Secretary of Transportation to study distracted driving, including cognitive distraction when driving and driver distraction impacts on young, inexperienced drivers.
- Require the Secretary to withhold 25% of a state's apportionment of certain federal-aid highway program funds for the fiscal year if the

> state has not enacted or is not enforcing a law that (1) prohibits, except in an emergency, an operator of a moving or idling motor vehicle on a public road from using a hand-held mobile device (other than a voice-activated, vehicle-integrated, or similar device, or a global positioning system [GPS] which is not vehicle-integrated); and (2) requires, upon conviction of a violation of such prohibition, the imposition of certain minimum penalties.

The bill was referred to the House Transportation and Infrastructure Committee Subcommittee on Highways and Transit. No further action has been taken.

Distracted Driving Prevention Act of 2011, H.R. 1772

Representative Eliot Engel introduced this bill on May 5, 2011. The bill would:

- Direct the Secretary of Transportation to make distracted driving prevention incentive grants for each fiscal year to states that enact laws that prohibit, with certain exceptions, and establish fines for texting and/or handheld cellphone use while driving.
- Require a state that receives a grant to allocate (1) at least 50% to educate and advertise to the public about the dangers of texting or using a cellphone while driving as well as to enforce the distracted driving law; and (2) up to 50% for other traffic safety improvement projects.
- Direct the Administrator of the NHTSA to administer a distracted driving national education program with at least two high-visibility education and advertising campaigns.
- Require the Secretary to establish a research program to study distracted driving by passenger and commercial vehicle drivers.
- Direct the FCC to report to Congress on existing and developing wireless communications technology that may be used to reduce problems associated with distracted driving.
- Require the Secretary to (1) issue regulations on the use of electronic or wireless devices, including cell phones and other distracting devices, by operators of commercial motor vehicles and school buses; and (2) prohibit their use in circumstances where it interferes with the driver's safe operation of the vehicles.

This bill was referred to the Committee on Transportation and Infrastructure and Committee on Energy and Commerce. No further action has been taken.

Consolidated and Further Continuing Appropriations Act, 2012, P.L. 112-55 (S. 1596) (Originally Transportation, Housing and Urban Development, and Related Agencies Appropriations Act, 2012 (H.R. 2112))

P.L. 112-55 directs the NHTSA and the Centers for Disease Control to report on the extent to which electronic devices can be causally linked to the reported rise in fatal accidents or injuries involving distracted driving, as well as the impact distracted driving prevention laws and enforcement actions can have on motorist behavior.

SMS Spam

The CAN-SPAM Act was and is intended to curb the amount of spam that consumers receive in their e-mail accounts. At the time the act was being considered in 2003, text messaging was in its infancy as a service. As discussed above, SMS messaging is not the same as messaging that uses a mobile phone's associated e-mail address (e.g., 2025551212@carrier.com). At this time, only the latter type of message is covered by CAN-SPAM; messages that are sent "phone-to-phone" through the SMSC are not.

There is no evident reason for messages that appear the same to a user and have the same effect on a user (generally, annoyance) to be treated differently under CAN-SPAM (Controlling the Assault of Non-Solicited Pornography and Marketing Act, P.L. 108-187). Resolving this discrepancy in the treatment of these two types of messages would require a change to the statute.

Inability of Consumers to Disable Text Messaging

In the past, some mobile service customers had expressed frustration to their congressional representatives about unwanted text messages and the inability to selectively block or completely disable text messaging on their phones. Carriers generally offer a range of text messaging packages, for example, 500 messages for $10, but some customers simply do not use text messaging and, therefore, pay a small fee every time they receive a message.

In December 2007, a class-action lawsuit was filed against T-Mobile in this matter.[6] At that time, most carriers already offered some form of text blocking to their customers. The *Appendix* contains information from that article that may be helpful to consumers.[7] In August 2008, T-Mobile began allowing text blocking. At this time, all major carriers provide this service.

Text Messaging Price Fixing

A class-action suit pending before the district court for the Northern District of Illinois accuses the four national wireless carriers—AT&T, Verizon, Sprint, and T-Mobile—of colluding to set the price of text messages. Plaintiffs allege that the defendants agreed, in 2005, to set the price for a single text message at $0.10 and then subsequently agreed to raise that price twice, first to $0.15 and then to $0.20. The plaintiffs allege that it is reasonable to infer that those price changes were the result of a prior agreement because the industry is concentrated, the defendants made identical price changes at about the same time, no defendant attempted to gain an advantage in the market by lowering its price (even though the price is far in excess of cost), and the defendants had ample opportunity to conspire through participation in a trade association, which frequently discussed matters related to text messaging. Plaintiffs' complaint also notes that the Antitrust Subcommittee of the Senate Judiciary Committee initiated an investigation into the price increases.

Defendants have denied that they engaged in any collusion and dispute the claim that the facts support any inference of unlawful conduct. Defendants have pointed out that most consumers do not purchase text messaging services on a message-by-message basis, but instead buy packaged plans that include hundreds of messages—or an unlimited number—each month. The structure and pricing of those bundled plans varies among the carriers—plaintiffs do not claim that defendants reached any agreement with respect to those plans—and the effective price of text messaging service has declined sharply as output has expanded dramatically. The wireless carriers have also argued that the increases in prices for single-use text messages occurred over the space of 6 to 11 months, a pattern that is more suggestive of "follow-the-leader" pricing—which is lawful—than any advance coordination. Defendants dispute that the wireless service providers ever discussed pricing at meetings of their trade association, and they note that neither the Antitrust Subcommittee's investigation, nor a DOJ investigation initiated at the subcommittee's request, found any evidence of collusion among the carriers.

Carrier Blocking of Common Short Code Messages

In 2007, Verizon notified NARAL Pro-Choice America that it would not participate in its CSC program. NARAL does not charge for its messages and users may opt-in or opt-out as desired, but Verizon stated that it does not accept programs from any group "that seeks to promote an agenda or distribute content that, in its discretion, may be seen as controversial or unsavory to any of [its] users."[8]

This decision was immediately criticized by free-speech advocates, although communications scholars pointed out that the company most likely, from a legal standpoint, did have the right to refuse to participate in the program.[9] Since text messages are not carried over the traditional telephone network, such messages are not protected under common carrier regulation. The next day, Verizon changed its decision and is now participating in NARAL's CSC program, saying in a statement that the decision had been "an incorrect interpretation of a dusty internal policy" that "was designed to ward against communications such as anonymous hate messaging and adult materials sent to children." The policy had been developed "before text messaging protections such as spam filters adequately protected customers from unwanted messages."[10]

This issue highlights the difficulty in applying the current regulatory structure to new services. While mobile providers appear to have the legal right to determine what information is available through their CSC programs, Congress may wish to consider whether and how political and other speech might be better protected in those programs. However, there have been no recent cases in which carriers have blocked CSCs.

Deceptive and Misleading Common Short Code Programs

Many third-party content providers use the CSC program and bill the usage through the mobile service provider. For example, content providers can allow mobile device users to download content (e.g., ringtones) or participate in SMS-based "chat." While most of these content providers are legitimate businesses, others use deceptive tactics to gain customers and run up unexpected charges.[11]

For example, as reported by CBS News in February 2008, some customers have subscribed to monthly services without reading the "fine print" and find that the charge is often difficult to remove because it is an independent third party rather than the customer's mobile service provider.[12]

The Mobile Marketing Association has developed "Consumer Best Practices Guidelines"[13] that it expects its members to follow. This code includes limiting subscription periods to one month, after which consumers must re-subscribe, and providing alerts to customers when their chat-related charges reach $25 increments. Although the best practices have not eliminated all misleading programs, over time the industry may bring its members into compliance. More clarity on industry efforts might allow policymakers an opportunity to assess the efficacy of those efforts.

Protecting Children from Inappropriate Content on Wireless Devices

As more mobile devices become equipped to access the web and additional content services are made available via CSCs, the risk of children downloading inappropriate content will likely increase. While carriers may follow a set of voluntary guidelines[14] to promote wireless safety for children, there is no way to guarantee that children will not be able to access inappropriate content by circumventing carrier-implemented safeguards.

The following types of material can be downloaded on many wireless devices, and may include content inappropriate for children:

- Images, such as background "wallpaper" for the phone screen.
- Games, including some games that are also available for gaming systems.
- Music and songs, including ring tones, ringback tones, and downloads of full songs.
- Video, including certain television shows, movies, and music videos, as well as video programming specially made for, and only available on, wireless devices.[15]

The wireless industry is working to ensure that children do not access inappropriate information over their wireless devices, but there is no definitive research on the success of these efforts. Whether current efforts to protect children from inappropriate content over wireless devices may be an issue of interest to policymakers.

"Sexting"

Sexting is a term coined by the media that generally refers to youth writing sexually explicit messages, taking sexually explicit photos of themselves or others in their peer group, and transmitting those photos and/or messages to their peers.[16] Sexting is not the same as a child sending a sexually explicit photo to an adult, however, the ramifications can be extremely serious because of how child pornography laws are written. In general, regardless of the age of the person who takes the photograph and/or sends it, that photograph is considered child pornography. This has led to situations in which underage girls have been charged with distributing child pornography and others in which teenagers have been required to register as sex offenders.

Although no federal charges have been brought in these types of cases yet, it is conceivable that they could. Congress may wish to consider whether children should be prosecuted under statutes intended to prosecute child predators and pornographers and whether, in certain cases, such prosecutions might be warranted.

A report conducted in 2008 by the National Campaign to Prevent Teen and Unplanned Pregnancy found that 20% of its respondents had indicated they had engaged in sexting. That study, however, had included 18- to 19-year-old adults, which significantly skewed the findings.[17] A more recent study published in January 2012 indicated that the problem is not as widespread as originally thought. The second study, published in *Pediatrics*, asked tweens and teens (ages 10- 17) if they had ever engaged in sexting. Of those asked, 2.5% of the respondents in the survey said "they had appeared in or created images that depicted themselves nude or nearly nude." But, when the researchers asked if the images "showed breasts, genitals or someone's bottom," only 1.3% said they had appeared in or created such images.[18]

The study also found that older teens are much more likely to appear in such images than younger children. Just under three-quarters of the 2.5% who appeared in or created nude or nearly nude images were 16 or 17. Only 6% of that 2.5% were between 10 and 12. About 7% of the youth had received a nude or nearly nude picture, but only 1% reported forwarding or posting the image. Of those who received such images, 56% were girls and 55% were 16 or 17. Just under 6% reported receiving sexually explicit images. The study found that some—but far from most—youth engaged in sexting were emotionally upset as a result. For example, "21 percent of respondents appearing in or creating images reported feeling very or extremely upset, embarrassed or afraid as a result, as did 25 percent of the youth receiving images."[19]

Legislation—112th Congress

In the 112th Congress, one bill has been introduced that includes a section on sexting.

Anti-Bullying and Harassment Act of 2011, H.R. 975

Representative Danny Davis introduced H.R. 975, the Anti-Bullying and Harassment Act of 2011, on March 9, 2011. The bill was referred to the Committee on Education and the Workforce Subcommittee on Early Childhood, Elementary, and Secondary Education on March 21, 2011. The bill includes language that includes sexting in the definition of cyberbullying if the transmittal of a "nude picture" constitutes bullying[20] "that is undertaken, in whole or in part, through use of technology or electronic communications (including electronic mail, internet communications, instant messages, or facsimile communications) to transmit images, text, sounds, or other data."

Mobile Cyberbullying

"Cyberbullying," harassing communications sent, for example, via e-mail or text messages or through social networking sites such as Facebook or MySpace, is a growing problem. The issue made national headlines in November 2007 after the suicide of Megan Meier, a 13-year-old Missouri girl. In that case, the mother of a former friend of Megan's set up a fake MySpace page, pretending to be a boy who had just moved to the area and was home-schooled. Within a few weeks of becoming "friends" with "Josh," on October 15, 2006, the tone of his messages changed drastically, with "Josh" saying he no longer wanted to be friends with Megan, because "he" had heard that she had been mean to some of her friends. On October 16, 2006, Megan hanged herself in her closet.

Although, as in the case described above, much cyberbullying takes place in the "wired" world, more recently, these sorts of messages are being sent from and to mobile devices. Since many mobile devices are capable of performing the same tasks as computers, these messages are now being sent via mobile instant messaging, the mobile websites of social networking sites, and text messaging.

Legislation—112th Congress

In the 112th Congress, one bill has been introduced that includes a section on cyberbullying.

Anti-Bullying and Harassment Act of 2011, H.R. 975

Representative Danny Davis introduced H.R. 975, the Anti-Bullying and Harassment Act of 2011, on March 9, 2011. The bill was referred to the House Education and the Workforce Committee Subcommittee on Early Childhood, Elementary, and Secondary Education on March 21, 2011. The bill includes language defining cyberbullying as bullying[21] "that is undertaken, in whole or in part, through use of technology or electronic communications (including electronic mail, internet communications, instant messages, or facsimile communications) to transmit images, text, sounds, or other data." A number of other bills are aimed at bullying, in general.

Privacy of Text Messages[22]

Text messages are routinely used to conduct government business, both officially and unofficially. As a result employers,[23] litigants, newspapers, law enforcement,[24] and public interest groups are increasingly seeking access to the contents of such communications in order to shed light on the workings of government. Most states have laws "prohibiting public officials from discussing official business privately and failing to disclose information regarding the operations of government."[25] On the other hand, some contend that text messages should be treated as private because of the nature of the delivery platforms or technological devices employed to send text messages. States are increasingly disregarding the public nature of the employee's position and the property ownership of the device used to send e-mail and text messages, and instead defining public communication based on whether the individual is conducting the public's business. Thus, communication about the public's business, and records documenting the communication are a part of the public record. Because text messaging represents a relatively new form of electronic communication, federal and state courts,[26] legislatures,[27] and commissions[28] are considering access to text messages pursuant to Sunshine Laws, Open Meetings Laws, Public Records Acts, Freedom of Information Acts, and electronic surveillance laws.

In Detroit, MI, newspapers filed a Michigan Freedom of Information Act (FOIA) lawsuit against that city seeking disclosure of text messages sent by

Detroit elected officials on city-issued pagers that related to the city's $8.4 million settlement of two whistle-blower lawsuits brought by former Detroit police officers.[29] The city argued that disclosure of the text messages would violate the federal Stored Communications Act, which outlaws unlawful access to stored communications.[30] A public records directive issued by the city states that all electronic communications sent on city equipment "is not considered to be personal or private."[31] Although the newspapers obtained the text messages through an anonymous source, they continued to press for the release of additional information under public records law.[32] A court ruled part of the information the newspapers wanted was public, the *Free Press* published text messages related to the cover-up, and the mayor and chief of staff were charged with eight felonies and convicted.[33] In *Flagg v. City of Detroit*,[34] the district court held that a city's text messages satisfied the definition of "public records" under Michigan's Freedom of Information Act (FOIA)[35] because they captured communications among city officials or employees in the performance of an official function. For purposes of the discovery rule governing production of documents, the district court concluded that the city had "control" over any such "public records" in the possession of a third party service provider under contract with the city by virtue of its statutory obligation to maintain these records and make them available for examination or inspection.[36]

Courts also have examined whether the disclosure of text messages sent by employees on employer-issued pagers violates the privacy rights of employees, and whether such disclosure is barred by the Stored Communications Act (SCA).[37] The United States Supreme Court in *City of Ontario v. Quon* overturned a federal appellate court decision which had held that officials in the city of Ontario, CA, engaged in an unconstitutional search and seizure when they acquired and read the contents of messages sent to and from a city police officer's city-provided pager.[38] In *Quon v. Arch Wireless Operating Company*,[39] the court of appeals for the Ninth Circuit concluded that, under certain circumstances, an employee sending text messages from an employer's device has a reasonable expectation of privacy under the Fourth Amendment. The United States Supreme Court resolved the case by applying settled principles for determining when a search is reasonable.[40] In *City of Ontario v. Quon*, the Supreme Court held that officials had acted reasonably when they reviewed transcripts of messages sent to and from Sergeant Quon's city-issued pager in order to determine whether service limits on the pager's use should be increased. The Court assumed, without deciding, that Quon had a reasonable expectation of privacy for Fourth Amendment purposes, but found that the search of the transcripts was reasonable.

Courts also have begun exploring ways to apply open government laws to text messages. In Texas, a state judge ordered the city of Dallas to turn over e-mails and text messages sent by city officials from personal accounts and personal hand-held devices to conduct city business, and held that the e-mails and messages were subject to disclosure under the Texas Public Information Act.[41]

Using SMS to Support Law Enforcement and Emergency Response

In May 2011, the FCC and FEMA announced the implementation of a Personal Localized Alerting Network (PLAN). This program was previously called the Commercial Mobile Alert System (CMAS), which has been under development since April 2008 under rules developed by the FCC.

The PLAN system will be operational in Washington, DC, and New York City by the end of 2011 and in the rest of the country by April 2012. PLAN will deliver emergency text messages to the public during emergencies and natural disasters.[42]

The program was mandated by the Warning, Alert and Response Network Act that was signed into law in 2006.[43] Under this law, the FCC was required to develop plans for a commercial mobile-alert system through which wireless carriers would voluntarily transmit text messages sent out by the government. The FCC has divided the types of messages the government will send out to mobile-phone users into three broad categories:[44]

- Presidential Alerts deal with national emergencies and will take precedence over any other impending alerts
- Imminent Threat Alerts deal with emergencies that may pose an imminent risk to people's lives or well-being.
- Child Abduction Emergency/AMBER alerts will be related to missing or abducted children.

In addition, the FCC says that all subscribers with roaming agreements will receive timely alerts "provided the subscriber's mobile device is configured for and technically capable of receiving alert messages from the roamed upon network."[45]

The architecture adopted by the FCC calls for a centralized alert-aggregator where federal and state emergency-response agencies would send their warning messages to be authenticated and dispersed to the appropriate participating commercial mobile services. Noting FEMA's role in developing the proposal for the adopted architecture, the FCC recommended the agency as its first choice to serve as the alert aggregator and FEMA has accepted that role.

The FCC has issued a Second Report and Further Notice of Proposed Rulemaking;[46] an Order on Reconsideration and Erratum;[47] and a Third Report and Order.[48] Of particular note, in the Third Report and Order, the FCC—

- adopted notification requirements for wireless providers that elect not to participate, or to participate only in part, with respect to new and existing subscribers;
- adopted procedures by which wireless providers may elect to transmit emergency alerts and to withdraw such elections;
- adopted a rule governing the provision of alert opt-out capabilities for subscribers;
- allowed participating wireless providers to recover costs associated with the development and maintenance of equipment supporting the transmission of emergency alerts; and
- adopted a compliance timeline under which participating wireless providers must begin CMAS deployment.

The FCC continues to refine the rules for providing PLAN/CMAS. The most recent set of requirements is contained in the *Third Report and Order*, released August 7, 2008 (Docket No. 07-287). The WARN Act did not provide a mandatory deadline for the implementation of PLAN/CMAS.

The National Continuity Programs (NCP) Directorate, within the Federal Emergency Management Administration (FEMA), will take on the responsibility of acting as a gateway and aggregator of alerts for dissemination through PLAN/CMAS.[49] On December 7, 2009, FEMA and the FCC jointly announced that FEMA had adopted the CMAS Government Interface Design specifications. This triggered requirements in the *Third Report and Order* for wireless carriers that have agreed to participate in the PLAN/CMAS program to begin development and testing. The deadlines established by the FCC give these carriers until April 7, 2012, to provide

PLAN/CMAS alerts sent through the IPAWS gateway.[50] The four major wireless carriers will participate in the program.[51]

Congressional and Industry Response to SMS-Related Issues

The issues discussed in this report have prompted different levels of response from Congress and the wireless industry:

- Issues that are being addressed by industry, so policymakers may wish to wait and see how those efforts play out;
- Issues that have not risen to a level of priority in Congress, but would require statutory action to effect change; and
- Issues that have triggered a legislative response.

As wireless communications technologies, and the issues that accompany them, evolve over time, so likely will the approaches that industry and Congress will take to ensure consumer safety and satisfaction.

Appendix. Text Blocking with Selected Major Carriers—Information for Consumers

AT&T

Customers must log in at mymessages.wireless.att.com. Text-blocking and alias options are available under "Preferences." Messages from specific e-mail addresses or websites can also be blocked from this page.

Verizon Wireless

Customers must log in at vtext.com. Text blocking options are available under "Text Messaging"/"Preferences." Select "Text Blocking." Consumers may block text messages from email or from the web, including blocking specific addresses or websites.

Sprint

Customers must log in at http://www.sprint.com. Sprint does not offer auto-blocking, but consumers can block specific phone numbers and addresses. On the top navigation bar, select, "My Online Tools"/"Communication Tools"/"Text Messaging." On the Compose a Text Message page, under Text Messaging Options, select "Settings & Preferences." In the text box, customers can enter a phone number, e-mail address, or domain name to block.

T-Mobile

Customers must log in at *http://www.t-mobile.com* and select "Communication Tools." T-Mobile doesn't yet offer a "block text messages from the Internet" option. Customers can block all messages sent by e-mail, though, or permit only messages sent to the phone's e-mail address or alias, or create filters that block text messages containing certain phrases.[52]

End Notes

[1] For some alphabets, such as Chinese, the maximum SMS size is 70 characters.

[2] See https://www.usshortcodes.com/csc/search/publicsearchCSC.do?method=showVanity & group=all for examples of such codes.

[3] See http://www.americanidol.com/mobile/ for specific instructions.

[4] Official U.S. Government website for Distracted driving, http://www.distraction.gov/stats-and-facts/.

[5] Section 712 addresses distracted driving.

[6] RCR Wireless News, "Class Action Nails T-Mobile USA Over Texting Services," January 30, 2008, available online at http://www.rcrnews.com/apps/pbcs.dll/article?AID=/ 200801 30/FREE/927035123/1005/rss01.

[7] New York Times, "How to Block Cellphone Spam," by David Pogue, June 12, 2008, available online at http://www.nytimes.com/2008/06/12/technology/personaltech/12pogue-email. html.

[8] New York Times, "Verizon Blocks Messages of Abortion Rights Group," by Adam Liptak, September 27, 2007, available online at http://www.nytimes.com/2007/09/27/us/ 27 verizon.html.

[9] New York Times, "Verizon Blocks Messages of Abortion Rights Group," by Adam Liptak, September 27, 2007, available online at http://www.nytimes.com/2007/09/27/us/ 27verizon.html.

[10] New York Times, "Verizon Reverses Itself on Abortion Messages," by Adam Liptak, September 28, 2007, available online at http://www.nytimes.com/2007/09/28/business/ 28verizon.html.

[11] See Class Action Connect online at http://www.classactionconnect.com/cell_phone_issues/ category/complaints-inthe-news/ for examples of these types of complaints.

[12] CBS News, "Ringing Up Big Charges For 'Free' Tones," February 22, 2008, available online at http://www.cbsnews.com/stories/2008/02/22/eveningnews/main3867197.shtml.

[13] This document is available online at http://www.mmaglobal.com/bestpractices.pdf.

[14] CTIA—The Wireless Association® has voluntary guidelines for wireless carriers to use in classifying content that they provide directly over wireless handsets. These voluntary guidelines apply only to content that you purchase from your wireless carrier, either on a one-time use or download basis, or as part of a package with a monthly fee such as ring tones, wallpaper, games, music, video clips, or TV shows. Content that is generated or owned by a wireless user, such as text messages, instant messages, e-mail (through chat rooms, message boards, etc.) and picture mail is not included in the wireless carrier's content classification system. Also, content that is accessed by surfing the Internet on a wireless handset is not currently included in the classification system. The guidelines urge carriers to provide separate web filtering software for web browsing services. Wireless carriers choosing to follow these voluntary guidelines agree to use at least two content ratings: (1) Generally Accessible or available to consumers of all ages; and (2) Restricted or accessible only to those age 18 and older or to those younger than 18 years old, when specifically authorized by a parent or guardian. The Restricted ratings system generally is based on or uses criteria under existing ratings systems for movies, television, music, and games. CTIA Guidelines are available online at http://www.ctia.org/advocacy/policy_topics/topic.cfm/TID/36.

[15] FCC Consumer Fact Sheet, "Protecting Children from Adult Content on Wireless Devices," available online at http://www.fcc.gov/cgb/consumerfacts/protectingchildren.html.

[16] National Conference of State Legislatures, 2009 Legislation Related to "Sexting" http://www.ncsl.org/?tabid=17756.

[17] "Sexting Far Less Prevalent Than Previously Reported," CNET, December 5, 2011, http://news.cnet.com/8301-19518_3-57336423-238/sexting-far-less-prevalent-than-previously-reported/.

[18] The "nude or nearly nude" category included youth wearing underwear or bathing suits or even fully clothed but in sexy poses.

[19] "Sexting Far Less Prevalent Than Previously Reported," CNET, December 5, 2011, http://news.cnet.com/8301-19518_3-57336423-238/sexting-far-less-prevalent-than-previously-reported/.

[20] The term "bullying" is defined in the act as conduct, including conduct that is based on a student's actual or perceived identity with regard to race, color, national origin, gender identity, disability, sexual orientation, religion, or any other distinguishing characteristics that may be defined by a state or local educational agency that (1) is directed at one or more students; (2) substantially interferes with educational opportunities or educational programs of such students; and (3) adversely affects the ability of a student to participate in or benefit from the school's educational programs or activities by placing a student in reasonable fear of physical harm."

[21] The term "bullying" is defined in the act as conduct, including conduct that is based on a student's actual or perceived identity with regard to race, color, national origin, gender identity, disability, sexual orientation, religion, or any other distinguishing characteristics that may be defined by a State or local educational agency that (1) is directed at one or more students; (2) substantially interferes with educational opportunities or educational programs of such students; and (3) adversely affects the ability of a student to participate in or benefit from the school's educational programs or activities by placing a student in reasonable fear of physical harm."

[22] Gina Stevens, Legislative Attorney in the CRS American Law Division, wrote this section.

[23] Lavis, Amanda J., "Employers Cannot Get the Message: Text Messaging and Employee Privacy," 54 Vill. L. Rev. 516 (2009).

[24] Alyssa H. DaCunha, "Txts R Safe 4 2Day: Quon v. Arch Wireless and the Fourth Amendment Applied to Text Messages," 17 Geo. Mason L. Rev. 296 (Fall 2009).

[25] Cooper, Cheryl, "Sending the Wrong Message: Technology, Sunshine Law, and the Public Record in Florida," 39 Stetson L. Rev. 411, 413 (2009-2010).

[26] See Disclosure of Electronic Data Under State Public Records and Freedom of Information Acts, 54 A.L.R.6th 653.

[27] A new Freedom of Information Law became effective in New York on August 7, 2008, and includes provisions which reflect a recognition of advances in information technology, but does not include a provision on text messaging. N.Y. Pub. Off. Law §84 et seq. (McKinney 2009). For a summary of the amendments to the Freedom of Information Law, see http://www.dos.state.ny.us/coog/foilnews2.html. See also "Battle Over Public Information Expands," by Ledyard King, Federal Times, March 24, 2008, p. 14.

[28] The Florida Commission on Open Government Reform reviewed open government laws to recommend areas for legislative review and amendment, and found that while e-mail communications between public officials become part of the public record, text or instant messages "most likely" do not. Ibid. at 413.

[29] Detroit Free Press, Inc., et al. v. City of Detroit, No. 08-100214 CZ, Wayne County Circuit Court, MI, at http://info.detnews.com/2008/0307motiontocompel.pdf.

[30] 18 U.S.C. §2701-2708. Subject to certain exceptions, the Stored Communications Act (SCA), which is part of the Electronic Communications Privacy Act, bars "a person or entity providing an electronic communications service to the public" from knowingly divulging to any person or entity the contents of a communication while in electronic storage by that service." 18 U.S.C. §2702(a)(1).

[31] On June 26, 2000, Mayor Kilpatrick signed a "Directive for the Use of the City of Detroit's Electronic Communications System."

[32] A "public record" under the Michigan Freedom of Information Act is "a writing that is: (1) prepared; (2) owned; (3) used; (4) in the possession of; or (5) retained by a public body in the performance of an official function...." Mich. Comp. Laws Ann. §15.232(e).

[33] For a chronology of developments, see Reporters Committee for Freedom of the Press, at http://www.rcfp.org/ newsitems/index.php?key=121&op=keyword.

[34] 252 F.R.D. 346 (E.D. Mich. 2008) (applying Michigan law).

[35] Mich. Comp. Laws Ann. §15.232(e).

[36] Ibid. at 356.

[37] 18 U.S.C. §2701 et seq.

[38] City of Ontario v. Quon, 130 S.Ct. 2619 (2010), rev'g, Quon v. Arch Wireless Operating Co., Inc., 529 F.3d 892 (9th Cir. 2008), cert. granted sub nom., City of Ontario v. Quon (Doc. No. 08-1332), 130 S.Ct. 1011 (2009).

[39] 529 F. 3d 892 (9th Cir, 2008) re'hg. en banc denied, 554 F. 3d 769 (9th Cir. 2009).

[40] CRS Report R41344, Public Employees' Right to Privacy in Their Electronic Communications: City of Ontario v. Quon in the Supreme Court, by Charles Doyle.

[41] Jennifer LaFleur, Dallas: City Must Provide Messages From Officials' Personal Accounts, Dallas Morning News, October 30, 2007, available at http://www.dallasnews.com/ sharedcontent/dws/news/localnews/stories/DNemails_30met.ART0.State.Edition1.421befa. html.

[42] Federal Communications Commission, In the Matter of the Commercial Mobile Alert System, First Report and Order, FCC 08-99, PS Docket No. 07-287, April 9, 2008, available online at http://hraunfoss.fcc.gov/edocs_public/ attachmatch/FCC-08-99A1.pdf ("Commercial Mobile Alert System, First Report and Order"). See also, FCC Adopts Rules for Delivery of Commerical Mobile Alerts to the Public During Emergencies (FCC 08-99), April 9, 2008, available online at http://hraunfoss.fcc.gov/edocs_public/attachmatch/FCC-08-99A1.pdf. See also the FCC's Consumer Fact Sheet on CMAS at http://www.fcc.gov/ cgb/consumer facts/cmas.html.

[43] Warning, Alert, and Response Network Act, Title VI of the Security and Accountability for Every Port Act of 2006, P.L. 109-347, 120 Stat. 1884 (2006).

[44] Commercial Mobile Alert System, First Report and Order, paras. 26-32.

[45] Commercial Mobile Alert System, First Report and Order, para. 79.

[46] Federal Communications Commission, In the Matter of the Commercial Mobile Alert System, Second Report and Further Notice of Proposed Rulemaking , FCC 08-164, PS Docket No. 07-287, July 8, 2008, available online at http://hraunfoss.fcc.gov/edocs_public/ attach match/FCC-08-164A1.pdf.

[47] Federal Communications Commission, In the Matter of the Commercial Mobile Alert System, Order on Reconsideration and Erratum, FCC 08-166, PS Docket No. 07-287, July 15, 2008, available online at http://hraunfoss.fcc.gov/edocs_public/attachmatch/FCC-08-166A1.pdf.

[48] Federal Communications Commission, In the Matter of the Commercial Mobile Alert System, Third Report and Order, FCC 08-184, PS Docket No. 07-287, July 15, 2008, available online at http://hraunfoss.fcc.gov/edocs_public/ attachmatch/FCC-08-184A1.pdf.

[49] "Nationwide Emergency Mobile Telephone Alert System Soon to Be Realized," Press Release, U.S. House of Representatives, Committee on Homeland Security, May 30, 2008.

[50] FCC Public Notice, "FCC's Public Safety and Homeland Security Bureau Sets Timetable in Motion for Commercial Mobile Alert Service Providers to Develop a System that Will Deliver Alerts to Mobile Devices," December 7, 2009, at http://hraunfoss.fcc.gov/edocs_ public/attachmatch/DA-09-2556A1.pdf.

[51] Daily Report for Executives, "FCC, FEMA Initiate Testing Phase of Wireless Emergency Alert System," by Alexel Alexis, December 8, 2009.

[52] "How to Block Cellphone Spam," NYTimes.com, Pogue's Posts, June 12, 2008, available online at http://pogue.blogs.nytimes.com/2008/06/12/how-to-block-cellphone-spam/?scp= 1&sq=Text%20Blocking&st=cse.

INDEX

J

K

L

M

Q

R

S

T

U

V

W

Y

Z